T0135924

Augsburger Schriften zur Mathematik, Physik und Informatik
Band 29

Bibliografische Information der Deutschen Nationalbibliothek

Die Deutsche Nationalbibliothek verzeichnet diese Publikation in der Deutschen Nationalbibliografie; detaillierte bibliografische Daten sind im Internet über http://dnb.d-nb.de abrufbar.

ISBN 978-3-8325-4061-6
ISSN 1611-4256

Logos Verlag Berlin GmbH
Comeniushof, Gubener Str. 47,
10243 Berlin
Tel.: +49 030 42 85 10 90
Fax: +49 030 42 85 10 92
INTERNET: http://www.logos-verlag.de

Stokes Structure and Direct Image of Irregular Singular $\mathcal{D}$-Modules

Dissertation

zur Erlangung des akademischen Grades
Dr. rer. nat.

eingereicht an der
Mathematisch-Naturwissenschaftlich-Technischen Fakultät
der
Universität Augsburg

von

Hedwig Heizinger

Augsburg, Dezember 2014

1. Gutachter: Prof. Dr. Marco Hien
2. Gutachter: Prof. Dr. Marc Nieper-Wißkirchen
3. Gutachter: Prof. Dr. Luis Narváez-Macarro

Tag der mündlichen Prüfung: 21. Mai 2015

Contents

1 Introduction **5**

2 Preliminaries **7**

3 Stokes-filtered local system $(\mathcal{L}, \mathcal{L}_{\leq\psi})$ **9**

3.1 Formal decomposition 9

3.2 Stokes-filtered local system 12

3.3 Topological description of the stalks 15

3.3.1 Construction of a Resolution of Singularities 19

3.3.2 Topology of $(\widetilde{p \circ e})^{-1}(\vartheta)$ 22

3.3.3 Explicit Description of B_{ψ}^{ϑ} 26

3.3.4 Dimension of $\mathbb{H}_c^1\left(B_{\psi}^{\vartheta}, \mathcal{F}_{\psi}\right)$ 34

4 Explicit example of the determination of Stokes data **43**

4.1 Stokes-filtered local system 43

4.2 Stokes data associated to $\mathcal{L}$ 51

4.3 Explicit computation of the Stokes matrices 55

Bibliography **69**

1 Introduction

In the 80s, Kashiwara and Mebkhout independently proved one of the most important results in the theory of $\mathcal{D}$-modules: the Riemann-Hilbert correspondence. It associates a flat meromorphic connection with regular singularities to a local system or, in general, a regular singular holonomic $\mathcal{D}$-module to a perverse sheaf. For irregular singularities Deligne and Malgrange stated a Riemann-Hilbert correspondence in the one-dimensional case. They described an equivalence of categories between irregular singular meromorphic connections and *Stokes-filtered local systems* (or in general irregular singular $\mathcal{D}$-modules corresponding to Stokes perverse sheaves) [Mal91].

Recently there has been a lot of progress in proving an irregular Riemann-Hilbert correspondence in higher dimensions. Sabbah introduced the notion of *good* meromorphic connections and proved an equivalence of categories in this case [Sab13], which was extended to the general case by Mochizuki [Moc09] and Kedlaya [Ked10]. Furthermore d'Agnolo/Kashiwara proved an irregular Riemann-Hilbert correspondence for all dimensions using subanalytic sheaves [DK].

Nevertheless it is still difficult to describe the Stokes phenomenon for explicit situations and to calculate Stokes data concretely. In their recent article, Hien/Sabbah have developed a topological way to determine Stokes data of the Laplace transform of an elementary meromorphic connection [HS]. The techniques used by Hien/Sabbah can be adapted to other situations. Hence in this work we will present a topological view of the Stokes phenomenon for the direct image of an exponentially twisted meromorphic connection $\mathcal{M}$ in the complex manifold $X = \Delta \times \mathbb{P}^1$:

$$\mathcal{N} := \mathcal{H}^0 p_+(\mathcal{M} \otimes \mathcal{E}^{\frac{1}{y}})$$

(where Δ is a small open disc around 0, p will denote the projection to Δ, y will denote the coordinate of $\mathbb{P}^1$ in ∞). First we give a precise description of the situation we will consider (Chapter 2). In Chapter 3 we will take a closer look at the Stokes-filtered local system $(\mathcal{L}, \mathcal{L}_{\leq \psi})$, which is associated to the $\mathcal{D}_\Delta$-module $\mathcal{N}$ by the irregular Riemann-

Hilbert correspondence as mentioned above. We will reprove an isomorphism proved by Mochizuki,

$$\Omega : \mathcal{L}_{\leq\psi} \xrightarrow{\cong} \mathcal{H}^1 R\widetilde{p}_* \,\mathrm{DR}^{mod\, D}\left(\mathcal{M} \otimes \mathcal{E}^{\frac{1}{y}} \otimes \mathcal{E}^{-\psi}\right)$$

This alternative proof implies in particular a topological description for the right hand side of the above formula and therefore for $\mathcal{L}_{\leq\psi}$. In Chapter 4 we will use this topological perspective to present a way of determining Stokes matrices for an explicit example.

The Stokes phenomenon, as it will be examined in this work, is based on a huge amount of definitions and constructions. We will not restate all of them, since the focus of this thesis should lie on applying these constructions to a particular situation in order to be able to describe Stokes-filtered local systems and calculate Stokes data explicitly. Therefore we assume the reader to be familiar with the used definitions and notations respectively we will refer to adequate literature.

Acknowledgements: I want to thank my advisor Marco Hien for accompanying this thesis during the last years. Marco, apart from your excellent mathematical support and instruction, I can not imagine a different 'Doktorvater', who could have introduced me to the world of mathematical research in a more motivating and trusting manner.
And of course thanks a lot to my family, friends and colleagues. You have made this work proceed by discussing mathematical problems with me as well as by distracting me from theoretical brainwork from time to time.

2 Preliminaries

Let $X = \Delta \times \mathbb{P}^1$ be a complex manifold, where Δ denotes an open disc in $0 \in \mathbb{C}$ with coordinate t. We denote the coordinate of $\mathbb{P}^1$ in 0 by x and the coordinate in ∞ by $y = \frac{1}{x}$. Let $\mathcal{M}$ be a regular singular holonomic $\mathcal{D}_X$-module. We have the following projections:

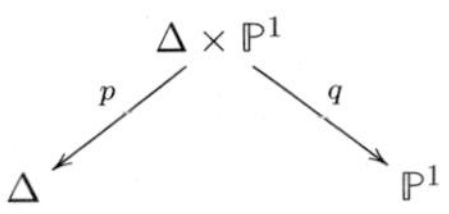

One should keep in mind, that locally at $(0, c) \in X$ $(c \neq \infty)$ we can choose coordinates (t, x) such that $p(t, x) = t$ and $q(t, x) = x$. Similarly locally at $(0, \infty) \in X$ we can choose coordinates (t, y) such that $p(t, y) = t$ and $q(t, y) = \frac{1}{y}$.

Let $\overline{D}$ denote the singular locus of $\mathcal{M}$, which consists of $\{0\} \times \mathbb{P}^1 =: D$, $\Delta \times \{\infty\}$ and some additional components. We will distinguish between the components

- $S_{i \subset I}$ $(I = \{1, \ldots, n\})$, which meet D in the point $(0, \infty)$ and
- $\widetilde{S}_{j \in J}$ $(J = \{1, \ldots, m\})$, which meet $\{0\} \times \mathbb{P}^1$ in some other point.

Furthermore we will require the following conditions on $\overline{D}$:

Assumption 2.1: *Locally in $(0, \infty)$ the irreducible components S_i of the divisor $\overline{D}$ achieve the following conditions:*

- *$S_i : \mu_i(t)\, y = t^{q_i}$, where μ_i is holomorphic and $\mu_i(0) \neq 0$.*
- *For $i \neq j$ either $q_i \neq q_j$ or $\mu_i(0) \neq \mu_j(0)$ holds.*

Assumption 2.2: *The irreducible components $\widetilde{S}_j$ intersect D in pairwise distinct points. Moreover we assume $\widetilde{S}_j$ to be smooth, i. e. locally around the intersection point they can be described as*

$$\widetilde{S}_j : \mu_j(t)\, x = t^{q_j}.$$

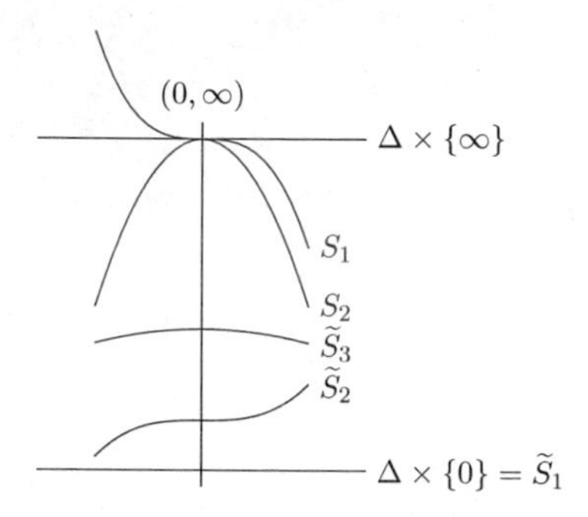

We want to examine the direct image of an exponentially twisted regular $\mathcal{D}_X$-module, namely $p_+(\mathcal{M} \otimes \mathcal{E}^q)$. This is a complex with $\mathcal{H}^k p_+(\mathcal{M} \otimes \mathcal{E}^q) = 0$ for $k \neq -1, 0$. Furthermore one can show that $\mathcal{H}^{-1} p_+(\mathcal{M} \otimes \mathcal{E}^q) = 0$ is supported in a single point (cf. [Sab08], p. 161). Thus the only interesting module to consider is $\mathcal{H}^0 p_+(\mathcal{M} \otimes \mathcal{E}^q)$. We will assume Δ small enough such that 0 is the only singularity of the $\mathcal{D}_\Delta$-module $\mathcal{H}^0 p_+(\mathcal{M} \otimes \mathcal{E}^q)$ and we denote its germ at 0 by

$$\mathcal{N} := \left(\mathcal{H}^0 p_+(\mathcal{M} \otimes \mathcal{E}^q)\right)_0 .$$

3 Stokes-filtered local system $(\mathcal{L}, \mathcal{L}_{\leq\psi})$

3.1 Formal decomposition

In her article *Formal structure of direct image of holonomic $\mathcal{D}$-modules of exponential type* Roucairol determined the exponential factors as well as the rank of the corresponding regular parts appearing in the formal decomposition of $\mathcal{N}$ [Rou07]. In our case this leads to the following

Theorem 3.1: *Let $\mathcal{M}$ be a regular singular $\mathcal{D}_X$-module with singular locus $\overline{D}$ which achieves the previous assumptions 2.1 and 2.2. Then $\hat{\mathcal{N}} := \left(\mathcal{H}^0 p_+ \left(\mathcal{M} \otimes \mathcal{E}^q\right)\right)_0^\wedge$ decomposes as*

$$\hat{\mathcal{N}} = R_0 \oplus \bigoplus_{i \in I} \left(R_i \otimes \mathcal{E}^{\psi_i(t)}\right),$$

where R_0, R_i $(i \in I)$ are regular singular $\mathcal{D}_\Delta$- modules and $\psi_i(t) = \mu_i(t)\, t^{-q_i}$. Moreover we have:

- *$rk(R_i) = \dim \Phi_{P_i}$ where Φ_{P_i} denotes the vanishing cycles of* $\mathrm{DR}(e^+\mathcal{M})$ *at the intersection point P_i of (a strict transform of) S_i with the exceptional divisor after a suitable blow-up map e.*
- *$rk(R_0) = \sum_{j \in J} \dim \Phi_{\widetilde{P}_j}$, where $\Phi_{\widetilde{P}_j}$ denotes the vanishing cycles of* $\mathrm{DR}(\mathcal{M})$ *at the intersection point $\widetilde{P}_j$ of $\widetilde{S}_j$ with D.*

Remark 3.2: 1. For details concerning the complex of vanishing cycles we refer to [Dim04], Ch. 4.2.

2. In Roucairol's article one will find $rk(R_i) = m_i$ where m_i is the multiplicity of the conormal space $T^*_{S_i}X$ in the characteristic cycle of $\mathcal{M}$. But this is equivalent to the statement above (cf. [Dim04], Prop. 4.3.20).

Proof: The formal decomposition and the statement about the rank of the R_i is exactly

Roucairol's theorem applied to our given situation (cf. [Rou07] and Remark 3.2 above). It remains to show

$$rk\,(R_0) = \sum_{j\in J} \dim\, \Phi_{\widetilde{P}_j}$$

We will use the same arguments as in Roucairol's proof.
R_0 is the regular part of $\mathcal{N}$. According to [Rou07] the rank of R_0 can be computed via the moderate nearby cycle functor ψ_t^{mod} (cf. [Sab08], p. 7).

$$rk\,(R_0) = \dim\left(\psi_t^{mod}\,(\mathcal{N})\right) = \chi\left(\psi_t^{mod}\,(\mathcal{N})\right)$$

We get the following equation (cf. [Rou07])

$$\begin{aligned}\psi_t^{mod}\,(\mathcal{N}) &= \psi_t^{mod}\left(\mathcal{H}^0 p_+\,(\mathcal{M}\otimes\mathcal{E}^q)_0\right)\\ &= R\Gamma\left(D, \mathrm{DR}\ \psi_p^{mod}\,(\mathcal{M}\otimes E^q)\,[+1]\right)\end{aligned} \tag{3.1}$$

We look at the De Rham complex pointwise. Let $P_\infty := (0,\infty)$ and P_j the intersection points of $\widetilde{S}_j$ with D. In local coordinates we have (cf. [Rou07], Lemma 3):

- $P_j = (0,c)$: $\psi_p^{mod}\,(\mathcal{M}\otimes\mathcal{E}^q)_{(0,c)} = \psi_t^{mod}\,(\mathcal{M}\otimes\mathcal{E}^x)_{(0,0)}$ and we get:

$$\chi\left(\mathrm{DR}\ \psi_t^{mod}\,(\mathcal{M}\otimes\mathcal{E}^x)_{(0,0)}\right) = \chi\left(\mathrm{DR}\ \psi_t^{mod}\,(\mathcal{M})_{(0,0)}\right) = r - \Phi_{\widetilde{P}_j}$$

- $Q = (0,c)\,(\neq P_j, P_\infty) : \psi_p^{mod}\,(\mathcal{M}\otimes\mathcal{E}^q)_{(0,c)} = \psi_t^{mod}\,(\mathcal{M}\otimes\mathcal{E}^x)_{(0,0)}$ and we get:

$$\chi\left(\mathrm{DR}\ \psi_t^{mod}\,(\mathcal{M}\otimes\mathcal{E}^x)_{(0,0)}\right) = \chi\left(\mathrm{DR}\ \psi_t^{mod}\,(\mathcal{M})_{(0,0)}\right) = r$$

- P_∞: $\psi_p^{mod}\,(\mathcal{M}\otimes\mathcal{E}^q)_{(0,\infty)} = \psi_t^{mod}\left(\mathcal{M}\otimes\mathcal{E}^{\frac{1}{y}}\right)_{(0,0)}$. There is an isomorphism

$$\mathcal{M}\otimes\mathcal{E}^{\frac{1}{y}} \cong e_+\left(e^+\mathcal{M}\otimes\mathcal{E}^{\frac{1}{y}\circ e}\right)$$

  for e a sequence of blow-ups (and therefore e a proper map) (cf. [Meb89], Prop. 7.4.5). As DR and ψ^{mod} commute with proper direct image we get

$$\mathrm{DR}\ \psi_t^{mod}\left(\mathcal{M}\otimes\mathcal{E}^{\frac{1}{y}}\right)_{(0,0)} = R\Gamma\left(E, \mathrm{DR}\ \psi_{t\circ e}^{mod}\left(e^+\mathcal{M}\otimes\mathcal{E}^{\frac{1}{y}\circ e}\right)\right)$$

  Here e denotes the blow-up of $(0,\infty)$ and $E = e^{-1}\,(0,\infty)$. E has local coordinates (u_1,v_1), (u_2,v_2) where $t = u_1 v_1, y = v_1$, respectively $t = u_2, y = u_2 v_2$. We can compute the Euler characteristic for DR $\psi^{mod}_{toe}\left(e^{+}\mathcal{M} \otimes \mathcal{E}^{\frac{1}{y}\circ e}\right)$ in each point of E (cf. [Rou07], Lemma 2):

- $Q = (0, v_2)$, $(v_2 \neq 0)$: $\chi\left(\text{DR } \psi^{mod}_{u_2}\left(e^{+}\mathcal{M} \otimes \mathcal{E}^{\frac{1}{u_2}}\right)_{(0,0)}\right) = 0$
- $Q = (0, v_2)$, $(v_2 = 0)$: $\chi\left(\text{DR } \psi^{mod}_{u_2}\left(e^{+}\mathcal{M} \otimes \mathcal{E}^{\frac{1}{u_2 v_2}}\right)_{(0,0)}\right) = 0$
- $P = (0, 0)$ (in coordinates (u_1, v_1)): $\chi\left(\text{DR } \psi^{mod}_{u_1 v_1}\left(e^{+}\mathcal{M} \otimes \mathcal{E}^{\frac{1}{v_1}}\right)_{(0,0)}\right) = -r$

Thus we get

$$\chi\left(\text{DR } \psi^{mod}_{t}\left(\mathcal{M} \otimes \mathcal{E}^{\frac{1}{y}}\right)_{(0,0)}\right) = \chi\left(R\Gamma\left(E, \text{DR } \psi^{mod}_{toe}\left(e^{+}\mathcal{M} \otimes \mathcal{E}^{\frac{1}{y}\circ e}\right)\right)\right) = -r.$$

Let $K^{\bullet} := \text{DR } \psi^{mod}_{p}(\mathcal{M} \otimes \mathcal{E}^{q})$. Denote $D^c := D \setminus \{P_1, \dots P_j, P_\infty\}$. We can determine the topological Euler characteristic of D^c: D^c consists of the projective line minus $J+1$ points. Since $\chi(\mathbb{P}^1) = 2$ and the Euler characteristic of a point is 1, we receive:

$$\chi(D^c) = -J + 1$$

Since $K^{\bullet}$ is constructible with respect to the stratification $\{D^c, P_j, P_\infty\}$ we can compute the Euler characteristic of $R\Gamma(D, K^{\bullet})$ with Mayer-Vietoris:

$$\begin{aligned}
\chi R\Gamma(D, K^{\bullet}) &= \chi(R\Gamma(D^c, K^{\bullet})) + \sum_j \chi(K^{\bullet})_{P_j} + \chi(K^{\bullet})_{P_\infty} \\
&= \chi(D^c) \cdot r + \sum_j (r - \widetilde{m}_j) - r \\
&= -J \cdot r + r + J \cdot r - \sum_j \widetilde{m}_j - r \\
&= -\sum_j \widetilde{m}_j
\end{aligned}$$

Because of the shift in (3.1) we get

$$\dim \psi^{mod}_t(\mathcal{N}) = \sum_j \Phi_{\widetilde{P}_j}$$

□

3.2 Stokes-filtered local system

Identify $\mathbb{S}^1 = \{\vartheta \mid \vartheta \in [0, 2\pi)\}$ and denote $\mathcal{P} := x^{-1}\mathbb{C}[x^{-1}]$. Let us recall the following definition of a Stokes-filtered local system.

Definition 3.3: Let $\mathcal{L}$ be a local system of $\mathbb{C}$-vector spaces on $\mathbb{S}^1$. We will call $(\mathcal{L}, \mathcal{L}_{\leq})$ a *Stokes-filtered local system*, if it is equipped with a family of subsheaves $\mathcal{L}_{\leq \phi}$ (indexed by $\phi \in \mathcal{P} := x^{-1}\mathbb{C}[x^{-1}]$) satisfying the following conditions:

1. For all $\vartheta \in \mathbb{S}^1$, the germs $\mathcal{L}_{\leq \phi, \vartheta}$ form an exhaustive increasing filtration of $\mathcal{L}_\vartheta$
2. $gr_\phi \mathcal{L} := \mathcal{L}_{\leq \phi}/\mathcal{L}_{< \phi}$ is a local system on $\mathbb{S}^1$ (where $\mathcal{L}_{<\phi,\vartheta} := \sum_{\psi <_\vartheta \phi} \mathcal{L}_{\leq \psi, \vartheta}$)
3. $\dim \mathcal{L}_{\leq \phi, \vartheta} = \sum_{\psi \leq_\vartheta \phi} \dim gr_\psi \mathcal{L}_\vartheta$

Assume that $\mathcal{M}$ (and therefore the direct image $\mathcal{N}$ defined above) is a meromorphic connection. Let $\mathcal{L}'$ denote the local system on Δ^* corresponding to the meromorphic connection $\mathcal{N}$. Moreover let

$$\pi : \widetilde{\Delta} \to \Delta, (r, e^{i\vartheta}) \mapsto r \cdot e^{i\vartheta}$$

be the real oriented blow-up of Δ in the singularity 0 and $j : \Delta^* \hookrightarrow \widetilde{\Delta}$. Consider $j_* \mathcal{L}'$ and restrict it to the boundary $\partial\Delta$. We get a local system on $\mathbb{S}^1 \cong \partial\widetilde{\Delta}$ and define

$$\mathcal{L} := j_* \mathcal{L}'_{|\mathbb{S}^1}$$

Moreover, consider the *moderate de Rham complex*

$$\mathrm{DR}^{mod\,0}(\mathcal{N}) : \{\mathcal{A}^{mod\,0} \otimes \pi^{-1}\mathcal{N} \to \mathcal{A}^{mod\,0} \otimes \pi^{-1}(\mathcal{N} \otimes \Omega^1)\}$$

and the *rapid decay de Rham complex*

$$\mathrm{DR}^{<0}(\mathcal{N}) : \{\mathcal{A}^{<0} \otimes \pi^{-1}\mathcal{N} \to \mathcal{A}^{<0} \otimes \pi^{-1}(\mathcal{N} \otimes \Omega^1)\}$$

where $\mathcal{A}^{mod\,0}$ (resp. $\mathcal{A}^{<0}$) denotes the sheaf of holomorphic functions on Δ^* having moderate growth (resp. rapid decay) along $\partial\widetilde{\Delta}$. For $\phi \in \mathcal{P}$ define

$$\mathcal{L}_{\leq \phi} := \mathcal{H}^0 \ \mathrm{DR}^{mod\,0}_{\partial\widetilde{\Delta}}(\mathcal{N} \otimes \mathcal{E}^{-\phi})$$

$$\mathcal{L}_{< \phi} := \mathcal{H}^0 \ \mathrm{DR}^{<0}_{\partial\widetilde{\Delta}}(\mathcal{N} \otimes \mathcal{E}^{-\phi}).$$

These are a subsheaves of $\mathcal{L}$. Actually the above construction defines a functor associating a germ of a $\mathcal{O}_\Delta(*0)$-connection to a Stokes-filtered local system $(\mathcal{L}, \mathcal{L}_\leq)$. Due to Deligne/Malgrange (cf. [Mal91]) we get:

Theorem 3.4: *The above functor is an equivalence of categories.*

By the Hukuhara-Turrittin-Theorem (cf. [Sab07], p. 109) the formal decomposition of $\mathcal{N}$ can be lifted locally on sectors to $\partial\widetilde{\Delta} = \mathbb{S}^1$. Thus to determine the filtration $\mathcal{L}_\leq$, it is enough to consider the set of exponential factors (respectively their polar part) appearing in the formal decomposition, since the moderate growth property of the solutions of an elementary connection $\mathcal{R} \otimes \mathcal{E}^\phi$ ($\phi \in P$) only depends on the asymptotical behavior of e^ϕ. We denote the set of exponential factors of the formal decomposition of $\mathcal{N}$ by

$$\mathcal{P}_\mathcal{N} := \{\psi_i \mid i \in I_0\}$$

whereby $I_0 := I \cup \{0\}$ and $\psi_0 := 0$.

Definition 3.5: For $\vartheta \in \mathbb{S}^1$ we define the following ordering on $\mathcal{P}$:

$$\phi \leq_\vartheta \psi :\Leftrightarrow e^{\phi-\psi} \in \mathcal{A}^{mod\,0}$$

Remark 3.6: 1. For $\psi_i, \psi_j \in \mathcal{P}_\mathcal{N}$ appearing in the formal decomposition of $\mathcal{N}$ we can determine $\vartheta \in \mathbb{S}^1$, such that $\psi_i \leq_\vartheta \psi_j$ (cf. [Sab13], ex. 1.4). Since

$$\psi_i \leq_\vartheta \psi_j \Leftrightarrow \psi_i - \psi_j \leq_\vartheta 0 \Leftrightarrow \mu_i(t)\, t^{-q_i} - \mu_j(t)\, t^{-q_j} \leq_\vartheta 0$$

we distinguish the following cases:

- $q_i < q_j$:

$$\psi_i \leq_\vartheta \psi_j \Leftrightarrow \left(\mu_i(t)\, t^{q_j-q_i} - \mu_j(t)\right) t^{-q_j} \leq_\vartheta 0 \Leftrightarrow arg(\;\;\mu_j(0)) - q_j\vartheta \in \left(\frac{\pi}{2}, \frac{3\pi}{2}\right)$$

- $q_i = q_j$:

$$\psi_i \leq_\vartheta \psi_j \Leftrightarrow \left(\mu_i(t) - \mu_j(t)\right) t^{-q_j} \leq_\vartheta 0 \Leftrightarrow arg\left(\mu_i(0) - \mu_j(0)\right) - q_j\vartheta \in \left(\frac{\pi}{2}, \frac{3\pi}{2}\right)$$

- $q_i > q_j$:

$$\psi_i \leq_\vartheta \psi_j \Leftrightarrow \left(\mu_i(t) - \mu_j(t)\, t^{q_i - q_j}\right) t^{-q_i} \leq_\vartheta 0 \Leftrightarrow arg\left(\mu_i(0)\right) - q_i\vartheta \in \left(\frac{\pi}{2}, \frac{3\pi}{2}\right)$$

There exists a finite set of $\vartheta \in \mathbb{S}^1$, where ψ_i and ψ_j are not comparable, i. e. neither $\psi_i \leq_\vartheta \psi_j$ nor $\psi_j \leq_\vartheta \psi_i$ holds. We call these angles the *Stokes directions* of (ψ_i, ψ_j).

2. For $\vartheta_0 \in \mathbb{S}^1$ not being a Stokes direction of any pair (ψ_i, ψ_j) in $\mathcal{P}_\mathcal{N}$ we get a total ordering of the ψ_is with respect to ϑ_0:

$$\psi_{\tilde{0}} <_{\vartheta_0} \ldots <_{\vartheta_0} \psi_{\tilde{n}}.$$

Corollary 3.7: *For $\vartheta \in \mathbb{S}^1$ we get the following statement:*

1. $\psi = 0$:

$$\dim\left(\mathcal{L}_{\leq 0}\right)_\vartheta = \sum_{j\in J} \Phi_{\widetilde{P}_j} + \sum_{\{i|\psi_i \leq_\vartheta 0\}} \Phi_{P_i}$$

2. $\psi \neq 0$, $\vartheta \in \left(\frac{\frac{\pi}{2}+arg(-\mu(0))}{q}, \frac{\frac{3\pi}{2}+arg(-\mu(0))}{q}\right)$ *mod* $\frac{2\pi}{q}$:

$$\dim\left(\mathcal{L}_{\leq \psi}\right)_\vartheta = \sum_{j\in J} \Phi_{\widetilde{P}_j} + \sum_{\{i|\psi_i \leq_\vartheta \psi\}} \Phi_{P_i}$$

3. $\psi \neq 0$, $\vartheta \in \left(\frac{-\frac{\pi}{2}+arg(-\mu(0))}{q}, \frac{\frac{\pi}{2}+arg(-\mu(0))}{q}\right)$ *mod* $\frac{2\pi}{q}$:

$$\dim\left(\mathcal{L}_{\leq \psi}\right)_\vartheta = \sum_{\{i|\psi_i \leq_\vartheta \psi\}} \Phi_{P_i}$$

Proof: This is a direct consequence of the formal decomposition and the fact that for $\psi \neq 0$ we have

$$0 \leq_\vartheta \psi \Leftrightarrow -\mu(t)\, t^{-q} \leq_\vartheta 0 \Leftrightarrow arg\left(-\mu(0)\right) - q\vartheta \in \left(\frac{\pi}{2}, \frac{3\pi}{2}\right)$$
$$\Leftrightarrow \vartheta \in \left(\frac{\frac{\pi}{2} + arg\left(-\mu(0)\right)}{q}, \frac{\frac{3\pi}{2} + arg\left(-\mu(0)\right)}{q}\right) mod\, \frac{2\pi}{q}$$

□

3.3 Topological description of the stalks

As before, we assume that $\mathcal{M}$ is a meromorphic connection with regular singularities along its divisor. Our aim is to determine the Stokes structure of $\mathcal{N}$ by using a topological point of view, i. e. we will develop a topological description of the Stokes-filtered local system $(\mathcal{L}, \mathcal{L}_{\leq})$. Therefore we will use the following theorem:

Theorem 3.8: *There is an isomorphism*

$$\Omega : \mathcal{L}_{\leq\psi} := \mathcal{H}^0 \operatorname{DR}^{mod\,0}_{\Delta} \left(\mathcal{N} \otimes \mathcal{E}^{-\psi}\right) \to \widetilde{\mathcal{L}}_{\leq\psi} := \mathcal{H}^1 R\widetilde{p}_* \operatorname{DR}^{mod\,D}_{\widetilde{X}(D)} \left(\mathcal{M} \otimes \mathcal{E}^{\frac{1}{y}} \otimes \mathcal{E}^{-\psi}\right)$$

Here $\widetilde{X}(D)$ denotes the real-oriented blow-up of X along the divisor component D, $\operatorname{DR}^{mod\,D}(\mathcal{M})$ denotes the moderate de Rham complex of a meromorphic connection $\mathcal{M}$ on X

$$\left\{\mathcal{A}^{mod\,D} \otimes \pi^{-1}\mathcal{M} \to \mathcal{A}^{mod\,D} \otimes \pi^{-1}(\mathcal{M} \otimes \Omega^1) \to \mathcal{A}^{mod\,D} \otimes \pi^{-1}(\mathcal{M} \otimes \Omega^2)\right\}$$

and $\widetilde{p} : \widetilde{X}(D) \to \widetilde{\Delta}$ corresponds to the projection p in the real oriented blow-up space $\widetilde{X}(D)$ along D. (See [Sab13], Ch. 8, for details on the definitions of real blow-up spaces and moderate de Rham complexes in higher dimensions.)

Theorem 3.8 is a special case of [Moc], Cor. 4.7.5. Nevertheless in the following chapters we will give an alternative proof. The existence of an injective morphism is given by the following Lemma due to Sabbah:

Lemma 3.9: *There is an injective morphism*

$$\Omega : \mathcal{L}_{\leq\psi} := \mathcal{H}^0 \operatorname{DR}^{mod\,0} \left(\mathcal{N} \otimes \mathcal{E}^{-\psi}\right) \to \widetilde{\mathcal{L}}_{\leq\psi} := \mathcal{H}^1 R\widetilde{p}_* \operatorname{DR}^{mod\,D} \left(\mathcal{M} \otimes \mathcal{E}^{\frac{1}{y}} \otimes \mathcal{E}^{-\psi}\right)$$

Proof: For a proof we refer to [Sab13], Lemma 13.10. □

Thus it remains to prove surjectivity. We will do this by showing that the dimensions at stalks level are equal, i.e. $\dim\left(\mathcal{L}_{\leq\psi}\right)_\vartheta = \dim\left(\widetilde{\mathcal{L}}_{\leq\psi}\right)_\vartheta$. As one will see, the proof provides a topological description of the stalks.

First let us examine the behavior of $\widetilde{\mathcal{L}}_{\leq\psi}$ with respect to birational maps.

Proposition 3.10: *Let $e : Z \to \Delta \times \mathbb{P}^1$ a birational map (i.e. a sequence of point blow-ups), $g(t,y) := \frac{1}{y} - \psi(t)$, $D_Z = e^{-1}(D)$. Then:*

$$\mathrm{DR}^{mod\, D}_{\widetilde{X}(D)}\left(\mathcal{M} \otimes \mathcal{E}^{g(t,y)}\right) \cong R\tilde{e}_* \, \mathrm{DR}^{mod\, D_Z}_{\widetilde{Z}(D_Z)}\left(e^+\mathcal{M} \otimes \mathcal{E}^{g(t,y) \circ e}\right)$$

Proof: As in the proof of Theorem 3.1 we will use the fact that

$$e_+\left(e^+\mathcal{M} \otimes \mathcal{E}^{g(t,y)\circ e}\right) \cong \mathcal{M} \otimes \mathcal{E}^{g(t,y)} .$$

We know that e is proper. Furthermore we assumed that $\mathcal{M}$ is a meromorphic connection with regular singularities along D, i.e. particularly that $\mathcal{M} \otimes \mathcal{E}^{g(t,y)}$ is a holonomic $\mathcal{D}_X$-module and localized along D. Thus we can apply [Sab13], Prop. 8.9:

$$\begin{aligned}\mathrm{DR}^{mod\, D}\left(\mathcal{M} \otimes \mathcal{E}^{g(t,y)}\right) &\cong \mathrm{DR}^{mod\, D}\left(e_+\left(e^+\mathcal{M} \otimes \mathcal{E}^{g(t,y)\circ e}\right)\right) \\ &\cong R\tilde{e}_* \, \mathrm{DR}^{mod\, D_Z}\left(e^+\mathcal{M} \otimes \mathcal{E}^{g(t,y)\circ e}\right)\end{aligned}$$

□

Proposition 3.11: *Denote $D' := D \cup (\Delta \times \{\infty\})$. Let $D'_Z := e^{-1}(D')$ and $\overline{D}_Z := e^{-1}(\overline{D})$. If $\overline{D}_Z$ is a normal crossing divisor, we have isomorphisms*

$$R\tilde{e}_* \, \mathrm{DR}^{mod\, D_Z}_{\widetilde{Z}(D_Z)}\left(e^+\mathcal{M} \otimes \mathcal{E}^{g(t,y)\circ e}\right) \cong R\tilde{e}_* \, \mathrm{DR}^{mod\, D'_Z}_{\widetilde{Z}(D'_Z)}\left(e^+\mathcal{M} \otimes \mathcal{E}^{g(t,y)\circ e}\right)$$

and

$$R\tilde{e}_* \, \mathrm{DR}^{mod\, D_Z}_{\widetilde{Z}(D_Z)}\left(e^+\mathcal{M} \otimes \mathcal{E}^{g(t,y)\circ e}\right) \cong R\tilde{e}_* \, \mathrm{DR}^{mod\, \overline{D}_Z}_{\widetilde{Z}(\overline{D}_Z)}\left(e^+\mathcal{M} \otimes \mathcal{E}^{g(t,y)\circ e}\right)$$

($\tilde{e}$ is interpreted as the map in the particular blow-up spaces)

Proof: Assume $\overline{D}_Z$ a normal crossing divisor and consider the identity map $Z \xrightarrow{Id} Z$, which obviously induces an isomorphism on $Z \setminus \overline{D}_Z \to Z \setminus \overline{D}_Z$. We obtain "partial" blow-up maps

$$\widetilde{Id}_1 : \widetilde{Z}(D'_Z) \to \widetilde{Z}(D_Z) \text{ and } \widetilde{Id}_2 : \widetilde{Z}\left(\overline{D}_Z\right) \to \widetilde{Z}(D_Z)$$

which induce the requested isomorphisms. These are variants of Prop. 8.9 in [Sab13] (see also [Sab13] Prop. 8.7 and Rem. 8.8).

□

As a direct consequence we get the following

Corollary 3.12:

$$\mathcal{H}^1\left(R\widetilde{p}_*\,\mathrm{DR}^{mod\,D}\left(\mathcal{M}\otimes\mathcal{E}^{g(t,y)}\right)\right)\cong\mathcal{H}^1\left(R\left(\widetilde{p\circ e}\right)_*\mathrm{DR}^{mod(\star)}\left(e^+\mathcal{M}\otimes\mathcal{E}^{g(t,y)\circ e}\right)\right)$$

where $\star = D_Z$ *(resp.* D'_Z*, resp.* $\overline{D}_Z$*)*

Proof: This follows from $R\widetilde{p}_*\circ R\widetilde{e}_* = R\left(\widetilde{p\circ e}\right)_*$.

□

As $(\mathcal{L}, \mathcal{L}_{\leq})$ was defined on $\partial\widetilde{\Delta}$ we will restrict our investigation to the boundaries of the relevant blow-up spaces, i. e. we consider $\partial\widetilde{\Delta}$, $\partial\widetilde{X}\,(D)$, $\partial\widetilde{Z}\,(D_Z)$, $\partial\widetilde{Z}\,(D'_Z)$ and $\partial\widetilde{Z}(\overline{D}_Z)$. Points beyond the boundary are holomorphic points and thus not relevant for the Stokes structure.

The diagram below gives a review over the relevant spaces and maps:

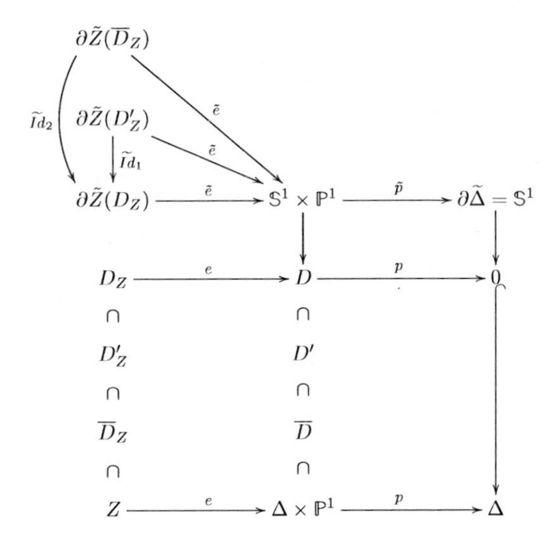

Lemma 3.13: *Let $\vartheta \in \mathbb{S}^1$. There is an isomorphism*

$$\left(\mathcal{H}^1\left(R\left(\widetilde{p \circ e}\right)_* \mathrm{DR}^{mod(D_Z)}\left(e^+\mathcal{M} \otimes \mathcal{E}^{g(t,y)\circ e}\right)\right)\right)_\vartheta \cong \mathbb{H}^1\left(\left(\widetilde{p \circ e}\right)^{-1}(\vartheta), \mathrm{DR}^{mod(D_Z)}\left(e^+\mathcal{M} \otimes \mathcal{E}^{g(t,y)\circ e}\right)\right)$$

The same holds for D'_Z and $\overline{D}_Z$ instead of D_Z.

Proof: $\widetilde{p \circ e}$ is proper, thus the claim follows by applying the proper base change theorem (cf. [Dim04], Th. 2.3.26, p. 41).

□

In the following we will consider

$$\mathcal{F}_\psi := \mathrm{DR}^{mod(D'_Z)}\left(e^+\mathcal{M} \otimes \mathcal{E}^{g(t,y)\circ e}\right)$$

on the fiber $\left(\widetilde{p \circ e}\right)^{-1}(\vartheta)$ in $\partial\widetilde{Z}\left(D'_Z\right)$, which in our case will be a 1-dimensional complex analytic space (See Chapter 3.2.2). Therefore we can assume $\mathcal{F}_\psi$ to be a perverse sheaf, i.e. a 2-term complex $\mathcal{F}_\psi : 0 \to \mathcal{F}^0 \to \mathcal{F}^1 \to 0$ with additional conditions on the cohomology sheaves $\mathcal{H}^0(\mathcal{F}_\psi)$ and $\mathcal{H}^1(\mathcal{F}_\psi)$. Obviously $\mathcal{H}^i(\mathcal{F}_\psi) = 0$ for $i \neq 0, 1$ (cf. [Dim04], Ex. 5.2.23, p. 139).
We can restrict $\mathcal{F}_\psi$ to the set

$$B_\psi^\vartheta := \left\{\zeta \in \left(\widetilde{p \circ e}\right)^{-1}(\vartheta) \mid \left(\mathcal{H}^\bullet(\mathcal{F}_\psi)\right)_\zeta \neq 0\right\},$$

which is an open subset of $\left(\widetilde{p \circ e}\right)^{-1}(\vartheta)$. We denote the open embedding by

$$\beta_\psi^\vartheta : B_\psi^\vartheta \hookrightarrow \left(\widetilde{p \circ e}\right)^{-1}(\vartheta).$$

Thus by interpreting $\mathcal{F}_\psi$ as a complex of sheaves on B_ψ^ϑ, we have to compute

$$\mathbb{H}^1\left(\left(\widetilde{p \circ e}\right)^{-1}(\vartheta), \beta_{\psi,!}^\vartheta \mathcal{F}_\psi\right) \cong \mathbb{H}_c^1\left(B_\psi^\vartheta, \mathcal{F}_\psi\right).$$

3.3.1 Construction of a Resolution of Singularities

In the following sections we will construct a suitable blow-up map e, such that we can describe $(\widetilde{p \circ e})^{-1}(\vartheta)$ and B_ψ^ϑ more concretely.

Lemma 3.14: *Let $g(t,y) = \frac{1}{y} - \psi(t)$. There exists a sequence of blow-up maps e such that*

1. *$g \circ e$ holomorphic or good, i. e.*

$$(g \circ e)(u,v) = \frac{1}{u^m v^n}\beta(u,v), \quad \textit{whereby } \beta \textit{ holomorphic and } \beta(0,v) \neq 0.$$

2. *For all i the strict transform of the divisor component S_i intersects D_Z in a unique point P_i.*

Proof: The divisor components S_i are given by $S_i : \mu_i(t)\, y = t^{q_i}$. Let $n := max\{q_i\}$. We distinguish two cases:

1. $\psi = 0$, i. e. $g(t,y) = \frac{1}{y}$. After n blow-ups in $(0,0)$ we get an exceptional divisor D_Z with local coordinates

$$t = u_k v_k, y = u_k^{k-1} v_k^k \text{ and } t = u_{\tilde{n}}, y = \tilde{u}_n^n \tilde{v}_n$$

and we have

- $(g \circ e)(u_1, v_1) = \frac{1}{v_1}$ is holomorphic for $v_1 \neq 0$ and good for $v_1 = 0$
- $(g \circ e)(u_k, v_k) = \frac{1}{u_k^{k-1} v_k^k}$ is good
- $(g \circ e)(\tilde{u}_n, \tilde{v}_n) = \frac{1}{\tilde{u}_n^n \tilde{v}_n}$ is good

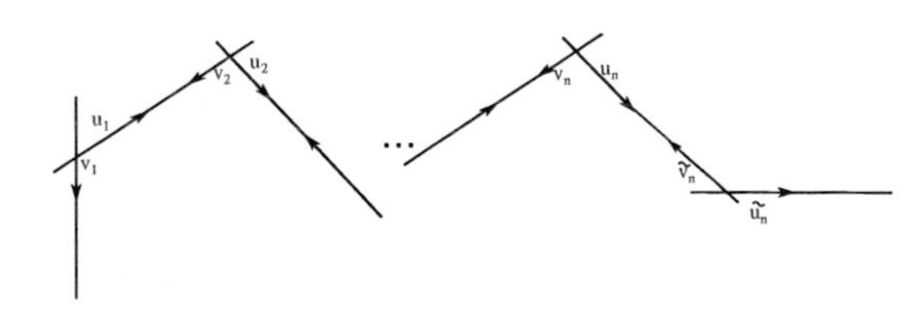

For computing the intersection point P_i of the strict transform of $S_i : \mu_i(t)\, y = t^{q_i}$ we consider again two different cases:

- $q_i < n$: Take the blow-up of order $q_i + 1$ and the corresponding local coordinates (u_{q_i+1}, v_{q_i+1}). The strict transform $\overline{S}_i$ given by the equation

$$\overline{S}_i : u_{q_i+1}^{q_i} v_{q_i+1}^{q_i} \left(1 - \mu_i \left(u_{q_i+1} v_{q_i+1}\right) v_{q_i+1}\right) = 0$$

intersects with D_Z uniquely in

$$P_i = \left(0, \frac{1}{\mu_i\left(0\right)}\right)$$

- $q_i = n$: We take the nth blow-up coordinates, whereby the strict transform of S_i is given by the equation

$$\overline{S}_i : \tilde{u}_n^n \left(1 - \mu_i \left(\tilde{u}_n\right) \tilde{v}_n\right) = 0$$

It intersects D_Z uniquely in

$$P_i = \left(0, \frac{1}{\mu_i\left(0\right)}\right)$$

2. $\psi \neq 0$: ψ is given by $\psi(t) = \mu(t) t^{-q}$, so $g(t, y) = \frac{t^q - \mu(t) y}{y t^q}$. In local coordinates $(g \circ e)$ is given by:
   - $q < n, k < n:$
     - $k \leq q : (g \circ e)(u_k, v_k) = \frac{u_k^{q-k+1} v_k^{q-k} - \mu(u_k v_k)}{u_k^q v_k^q}$
     - $k > q : (g \circ e)(u_k, v_k) = \frac{1 - \mu(u_k v_k) u_k^{k-1-q} v_k^{k-q}}{u_k^{k-1} v_k^k}$
   - $q < n, k = n : (g \circ e)(\tilde{u}_n, \tilde{v}_n) = \frac{1 - \mu(\tilde{u}_n) \tilde{u}_n^{n-q} \tilde{v}_n}{\tilde{u}_n^n \tilde{v}_n}$
   - $q = n, k < n : (g \circ e)(u_k, v_k) = \frac{u_k^{n-k+1} v_k^{n-k} - \mu(u_k v_k)}{u_k^q v_k^q}$
   - $q = n, k = n : (g \circ e)(\tilde{u}_n, \tilde{v}_n) = \frac{1 - \mu(\tilde{u}_n) \tilde{v}_n}{\tilde{u}_n^n \tilde{v}_n}$

   $(g \circ e)$ is good in every point except:
   - $q < n, k = q + 1 : P = \left(0, \frac{1}{\mu(0)}\right)$ with local coordinates (u_{q+1}, v_{q+1})
   - $q = n, k = n : P = \left(0, \frac{1}{\mu(0)}\right)$ with local coordinates $(\tilde{u}_n, \tilde{v}_n)$

   Let $q < n$. After changing coordinates $u' = u_k, v' = v_k - \frac{1}{\mu(u_k v_k)}$ and after q blow-ups in $(0, 0)$, we get:
   - $s \leq q : (g \circ e)(u'_s, v'_s) = \frac{-\mu(u_k v_k)}{{u'}_s^{q-s+1} {v'}_s^{q-s} \left({u'}_s^{s-1} {v'}_s^{s} + \frac{1}{\mu(u_k v_k)}\right)}$

- $s = q : (g \circ e)\left(\tilde{u}'_q, \tilde{v}'_q\right) = \frac{-\tilde{v}'_q\, \mu(u_k v_k)}{\tilde{u}'^q_q \tilde{v}'_q + \frac{1}{\mu(u_k v_k)}}$

Notice: For $u'_s = 0$ (just as for $\tilde{u}'_q = 0$) we have $\mu\left(u_k, v_k\right) = \mu\left(u'_s v'_s v_k\right) = \mu\left(0\right) \neq 0$. It follows that $(g \circ e)$ is good for every point of D_Z in local coordinates (u'_s, v'_s) and holomorphic for every point of D_Z in local coordinates $\left(\tilde{u}'_q, \tilde{v}'_q\right)$

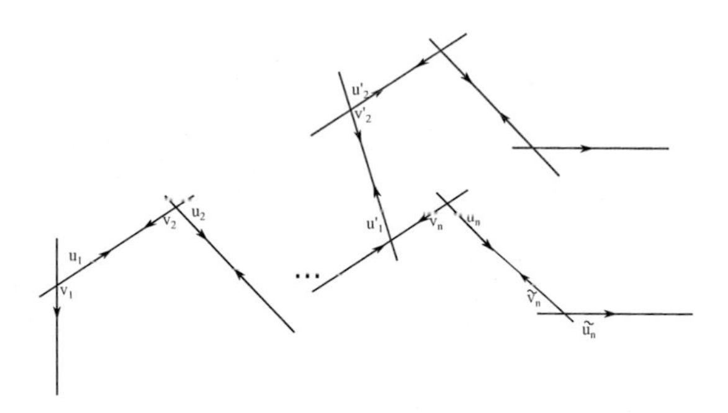

Let $q = n$: Change of coordinates $u' = \tilde{u}_n, v' = \tilde{v}_n - \frac{1}{\mu(\tilde{u}_n)}$ and n blow-ups in $(0, 0)$:

- $s \leq n : (g \circ e)\left(u'_s, v'_s\right) = \frac{-\mu(u'_s v'_s)}{u'^{n-s+1}_s v'^{n-s}_s \left(u'^{s-1}_s v'^s_s + \frac{1}{\mu(u'_s v'_s)}\right)}$
- $s = n : (g \circ e)\left(\tilde{u}'_n, \tilde{v}'_n\right) = \frac{-\tilde{v}'_n\, \mu(\tilde{u}'_n)}{\left(\tilde{u}'^n_n \tilde{v}'_n + \frac{1}{\mu(\tilde{u}'_n)}\right)}$

As before $(g \circ e)$ is good for every point of D_Z in local coordinates (u'_s, v'_s) and holomorphic for every point of D_Z in local coordinates $(\tilde{u}'_n, \tilde{v}'_n)$.

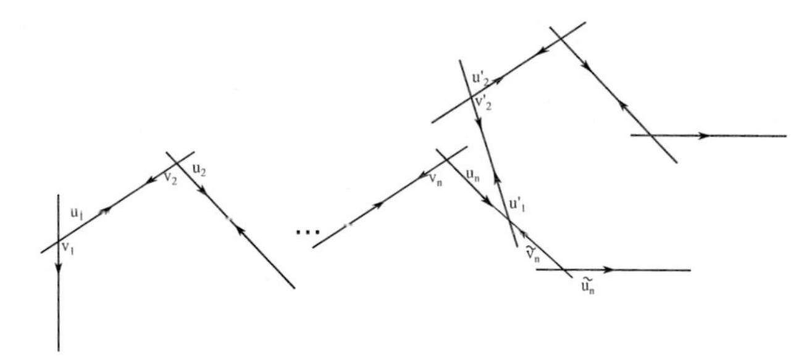

Now we compute the intersection points P_i of the strict transforms of the divisor components S_i with D_Z, where $S_i : \mu_i\left(t\right) y = t^{q_i}$.
First we consider the case $q_i \neq q$ or $\mu_i\left(0\right) \neq \mu\left(0\right)$.

- $q_i < n : \overline{S}_i : u_{q_i+1}^{q_i} v_{q_i+1}^{q_i} \left(1 - \mu_i \left(u_{q_i+1} v_{q_i+1}\right) v_{q_i+1}\right) = 0$ and thus

$$P_i = \left(0, \frac{1}{\mu_i(0)}\right)$$

- $q_i = n : \overline{S}_i : \tilde{u}_n^n \left(1 - \mu_i(\tilde{u}_n) \tilde{v}_n\right) = 0$ and

$$P_i = \left(0, \frac{1}{\mu_i(0)}\right)$$

Let $q_i = q$ and $\mu_i(0) = \mu(0)$, i. e. we consider $S : \mu(t) y = t^q$ (Notice that S corresponds to our given ψ!)

- $q < n$, i.e. $k := q + 1 \neq n$: $\overline{S} : u_k^{k-1} v_k^{k-1} \left(1 - \mu_i(u_k v_k) v_k\right) = 0$ and by coordinate transform we have: $\overline{S} : u_k^{k-1} v_k^{k-1} v' = 0$. As $v' = \tilde{u}'_q \tilde{v}'_q$ we get the unique intersection point

$$P = (0, 0)$$

- $q = n$: In the same way we get the intersection point

$$P = (0, 0)$$

□

3.3.2 Topology of $\left(\widetilde{p \circ e}\right)^{-1}(\vartheta)$

D'_Z is a normal crossing divisor, so locally at a crossing point D'_Z has the form $\{uv = 0\}$ and at a smooth point D'_Z has the form $\{u = 0\}$. Thus we can describe $\partial \widetilde{Z}(D'_Z)$ in local coordinates:

- real blow-up with respect to $\{u = 0\}$: $\zeta = (0, \theta_u, |v|, \theta_v)$ (with $v = |v| \cdot e^{i\theta_v}$)
- real blow-up with respect to $\{uv = 0\}$: $\zeta = (0, \theta_u, 0, \theta_v)$

Now we take the fiber of $\vartheta \in \mathbb{S}^1$, i. e. we consider $\left(\widetilde{p \circ e}\right)^{-1}(\vartheta)$. In local coordinates, every $\zeta \in \left(\widetilde{p \circ e}\right)^{-1}(\vartheta)$ has one of the following presentations:

- $\zeta = (0, \theta_{u_k}, |v_k|, \theta_{v_k})$ with $\theta_{u_k} + \theta_{v_k} = \vartheta$, $|v_k| \in [0, \infty)$
- $\zeta = (0, \vartheta, |\tilde{v}_n|, \theta_{\tilde{v}_n})$, $|\tilde{v}_n| \in [0, \infty)$

- $q < n$
  - $\zeta = (0, \theta_{u'_s}, |v'_s|, \theta_{v'_s})$ with $\theta_{u'_s} + \theta_{v'_s} = \theta_{u_k} = \vartheta - \theta_{v_k} = \vartheta - arg\left(\frac{1}{\mu(0)}\right)$, $|v'_s| \in [0, \infty)$
  - $\zeta = \left(0, \theta_{\tilde{u}'_q}, \left|\tilde{v}'_q\right|, \theta_{\tilde{v}'_q}\right)$ with $\theta_{\tilde{u}'_q} = \theta_{u_k} = \vartheta - \theta_{v_k} = \vartheta - arg\left(\frac{1}{\mu(0)}\right)$, $|\tilde{v}'_q| \in (0, \infty)$
  - $P := \zeta = \left(0, \theta_{\tilde{u}'_q}, 0\right)$ with $\theta_{\tilde{u}'_q} = \theta_{u_k} = \vartheta - \theta_{v_k} = \vartheta - arg\left(\frac{1}{\mu(0)}\right)$
- $q = n$
  - $\zeta = (0, \theta_{u'_s}, |v'_s|, \theta_{v'_s})$ with $\theta_{u'_s} + \theta_{v'_s} = \theta_{\tilde{u}_n} = \vartheta$, $|v'_s| \in [0, \infty)$
  - $\zeta = (0, \vartheta, |\tilde{v}'_n|, \theta_{\tilde{v}'_n})$, $|\tilde{v}'_n| \in (0, \infty)$
  - $P := \zeta = (0, \vartheta, 0)$ in local coordinates $(|\tilde{u}'_n|, \theta_{\tilde{u}'_n}, v'_{\tilde{n}})$
- $\zeta_0 = (0, \vartheta, 0)$. This is the point corresponding to the smooth point $(t, x) = (0, 0) \in D_Z$.

For every fixed $|v|$ we have a bijection $\{(0, \theta_u, |v|, \theta_v)\} \leftrightarrow \mathbb{S}^1$, thus we can interpret $(\widetilde{p \circ e})^{-1}(\vartheta)$ as a system of pipes.

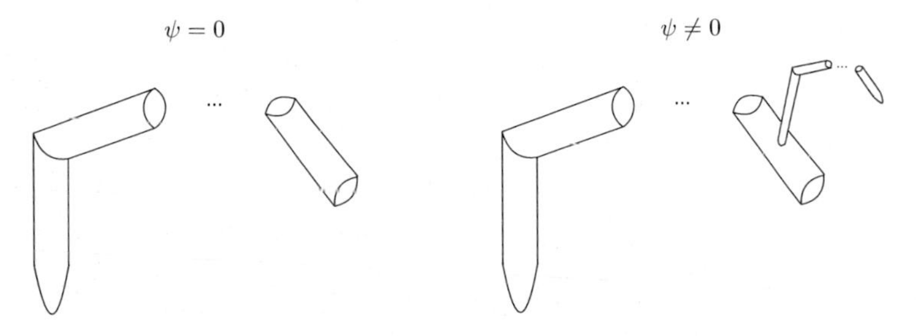

Furthermore these are homeomorphic to a disc, where the inlaying circles correspond to the changes of coordinates.

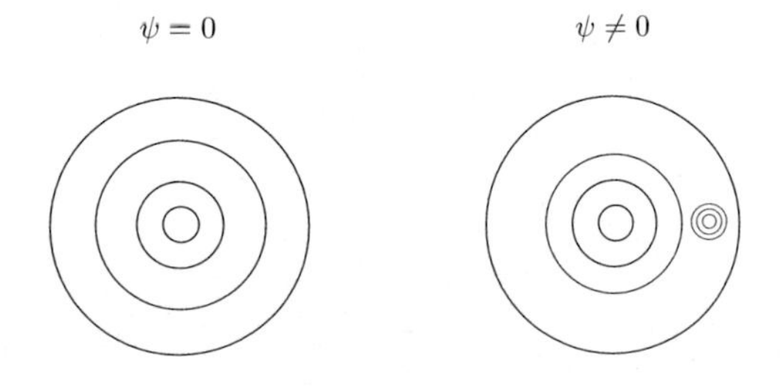

Lemma 3.15: *For all i the strict transform of the divisor component S_i intersects the real blow-up divisor $(\widetilde{p \circ e})^{-1}(\vartheta)$ in a unique point P_i.*

Proof:

1. $\psi = 0$: S_i is given by $S_i : \mu_i(t)\, y = t^{q_i}$.
   - $q_i < n$: According to the proof of Lemma 3.14 the strict transform

     $$\overline{S}_i : u_{q_i+1}^{q_i} v_{q_i+1}^{q_i} \left(1 - \mu_i\left(u_{q_i+1} v_{q_i+1}\right) v_{q_i+1}\right) = 0$$

     intersects with D_Z uniquely in $(u_{q_i+1}, v_{q_i+1}) = \left(0, \frac{1}{\mu_i(0)}\right)$. Taking the real-blow-up coordinates we get a unique intersection point

     $$P_i = \left(0, \theta_{u_{q_i+1}}, \left|\frac{1}{\mu_i(0)}\right|, \theta_{v_{q_i+1}}\right)$$

     where

     - $\theta_{v_{q_i+1}}$ is given by $\theta_{v_{q_i+1}} = arg\left(\frac{1}{\mu_i(0)}\right)$
     - $\theta_{u_{q_i+1}}$ is given by $\theta_{u_{q_i+1}} + \theta_{v_{q_i+1}} = \vartheta$
   - $q_i = n$: According to the proof of Lemma 3.14 the strict transform

     $$\overline{S}_i : \tilde{u}_n^n \left(1 - \mu_i(\tilde{u}_n)\, \tilde{v}_n\right) = 0$$

     intersects in $(\tilde{u}_n, \tilde{v}_n) = \left(0, \frac{1}{\mu_i(0)}\right)$. Thus the unique intersection point with $(\widetilde{p \circ e})^{-1}(\vartheta)$ is given by

     $$P_i = \left(0, \vartheta, \left|\frac{1}{\mu_i(0)}\right|, \theta_{\tilde{v}_n}\right)$$

     with $\theta_{\tilde{v}_n} = arg\left(\frac{1}{\mu_i(0)}\right)$
2. $\psi \neq 0$, where ψ is given by $\psi(t) = \mu(t)\, t^{-q}$:
   - $q_i \neq q$ or $\mu_i(0) \neq \mu(0)$:
     - $q_i < n$: $\overline{S}_i$: $u_{q_i+1}^{q_i} v_{q_i+1}^{q_i} \left(1 - \mu_i\left(u_{q_i+1} v_{q_i+1}\right) v_{q_i+1}\right) = 0$ intersects in $\left(0, \frac{1}{\mu_i(0)}\right)$ and thus

       $$P_i = \left(0, \theta_{u_{q_i+1}}, \left|\frac{1}{\mu_i(0)}\right|, \theta_{v_{q_i+1}}\right)$$

where

 * $\theta_{v_{q_i+1}}$ is given by $\theta_{v_{q_i+1}} = arg\left(\frac{1}{\mu_i(0)}\right)$
 * $\theta_{u_{q_i+1}}$ is given by $\theta_{u_{q_i+1}} + \theta_{v_{q_i+1}} = \vartheta$

- $q_i = n : \overline{S}_i : \tilde{u}_n^n \left(1 - \mu_i\left(\tilde{u}_n\right)\tilde{v}_n\right) = 0$ intersects in $\left(0, \frac{1}{\mu_i(0)}\right)$ and

$$P_i = \left(0, \vartheta, \left|\frac{1}{\mu_i\left(0\right)}\right|, \theta_{\tilde{v}_n}\right)$$

with $\theta_{\tilde{v}_n} = arg\left(\frac{1}{\mu_i(0)}\right)$

- $q_i = q$ and $\mu_i\left(0\right) = \mu\left(0\right)$:
  - $q < n$, i.e. $k := q + 1 \neq n$: $\overline{S} : u_k^{k-1} v_k^{k-1} \tilde{u}_q' \tilde{v}_q' = 0$ intersects in $(0, 0)$ and we get

$$P = \left(0, \theta_{\tilde{u}_q'}, 0\right)$$

  - $q = n$: In the same way we get the intersection point

$$P = \left(0, \theta_{\tilde{u}_n'}, 0\right)$$

□

Remark 3.16: 1. Also the irreducible components $\widetilde{S}_j$ intersect $\left(\widetilde{p \circ e}\right)^{-1}(\vartheta)$ in distinct points $\widetilde{P}_j$. This follows directly from Assumption 2.2.

2. According to Assumption 2.1 we have $\left(q_i \neq q_j \text{ or } \mu_i\left(0\right) \neq \mu_j\left(0\right) \text{ for } i \neq j\right)$. This induces $P_i \neq P_j$.

3. Every intersection point $P_i, \widetilde{P}_j$ may be interpreted as a 'leak' in the system of pipes $\left(\widetilde{p \circ e}\right)^{-1}(\vartheta)$. Thus topologically we can think of $\left(\widetilde{p \circ e}\right)^{-1}(\vartheta)$ as a disc with singularities, which come from the intersection with $\widetilde{S}_j$ and $\overline{S_i}$.

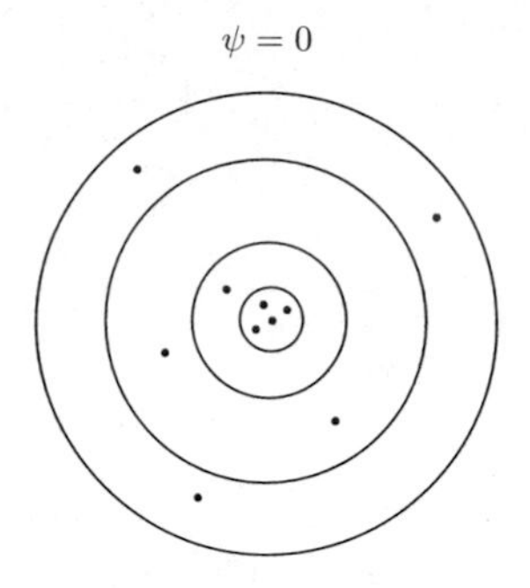

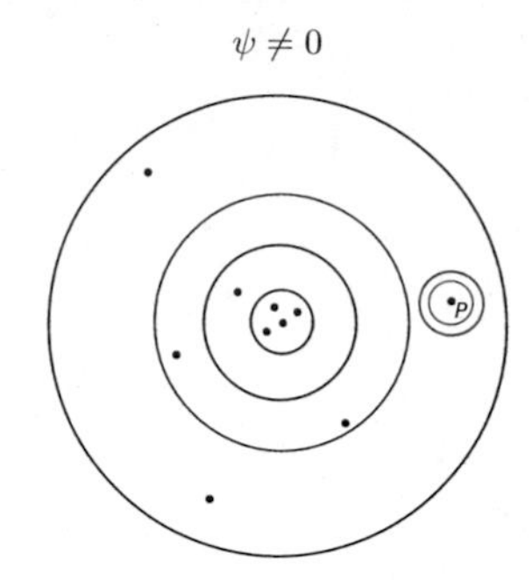

3.3.3 Explicit Description of B_ψ^ϑ

Remember the definition of

$$B_\psi^\vartheta := \left\{\zeta \in \left(\widetilde{p \circ e}\right)^{-1}(\vartheta) \mid (\mathcal{F}_\psi)_\zeta \neq 0\right\}$$

where $\mathcal{F}_\psi := \mathrm{DR}^{mod\, D'_Z}\left(e^+\mathcal{M} \otimes \mathcal{E}^{g(t,y)\circ e}\right)$. Obviously we have:

$$\zeta \in B_\psi^\vartheta \Leftrightarrow (\mathcal{F}_\psi)_\zeta \neq 0 \Leftrightarrow e^{(g\circ e)} \in \mathcal{A}_\vartheta^{mod\, D'_Z} \text{ near } \zeta$$

(The second equivalence follows by considering $\mathrm{DR}^{mod\, \overline{D}_Z}\left(e^+\mathcal{M} \otimes \mathcal{E}^{g(t,y)\circ e}\right)$, whereof we know that it has cohomology in degree 0 at most (cf. Proposition 3.11 and Lemma 4.3).)

Thus we need to take a closer look at the exponent $g \circ e$.

Lemma 3.17: *Let u, v local coordinates of the divisor D_Z, such that $f(u,v)$ holomorphic or good, i. e.*

$$f(u,v) = \frac{1}{u^m v^n}\beta(u,v), \quad \text{whereby } \beta \text{ holomorphic and } \beta(0,v) \neq 0.$$

Then $e^{f(u,v)} \in \mathcal{A}_\vartheta^{mod\, D_Z}$ around a given point $\zeta \in \left(\widetilde{p \circ e}\right)^{-1}(\vartheta)$ if and only if

$$f \text{ holomorphic in } \zeta \text{ or } arg\left(\beta(0,v)\right) - m\theta_u - n\theta_v \in \left(\frac{\pi}{2}, \frac{3\pi}{2}\right) \; mod\, 2\pi.$$

Proof: We have $Re(f(u,v)) = |f(u,v)| \, cos(arg(f(u,v)))$, because

$$\begin{aligned} f(u,v) &= |f(u,v)| \cdot e^{iarg(f(u,v))} \\ &= |f(u,v)| \cdot (cos(arg(f(u,v))) - isin(arg(f(u,v)))) \\ &= |f(u,v)| \, cos(arg(f(u,v))) - i\,|f(u,v)| \, sin(arg(f(u,v))) \end{aligned}$$

We know that $e^{f(u,v)} \in \mathcal{A}^{mod\, D'_Z} \Leftrightarrow Re(f(u,v)) < 0$ or f holomorphic. Thus assume f is not holomorphic. Then

$$\begin{aligned} e^{f(u,v)} \in \mathcal{A}^{mod\, D'_Z} &\Leftrightarrow |f(u,v)| \, cos(arg(g(u,v))) < 0 \\ &\Leftrightarrow cos(arg(f(u,v))) < 0 \\ &\Leftrightarrow arg(f(u,v)) \in \left(\frac{\pi}{2}, \frac{3\pi}{2}\right) \, mod\, 2\pi \\ &\Leftrightarrow arg(\beta(u,v)) - m\theta_u - n\theta_v \in \left(\frac{\pi}{2}, \frac{3\pi}{2}\right) \, mod\, 2\pi \end{aligned}$$

By plugging in $\zeta = (0, \theta_u, |v|, \theta_v)$ we get the requested equivalence.

□

Lemma 3.18: *Let $S_i : \mu_i(t)\, y = t^{q_i}$, $\psi_i(t) = \mu_i(t)\, t^{-q_i}$ with $\mu_i(0) \neq 0$ and P_i the corresponding intersection point. Then we have:*

$$P_i \in B_\psi^\vartheta \Leftrightarrow \psi_i \leq_\vartheta \psi$$

Proof: As before we denote $g(t,y) = \frac{1}{y} - \psi(t)$. Remark that $\mu_i(0) \neq 0$ particularly implies $\psi_i \neq 0$.
Let first $\psi = 0$: P_i is given by

$$P_i = \left(0, \theta_{u_{q_i+1}}, \left|\frac{1}{\mu_i(0)}\right|, arg\left(\frac{1}{\mu_i(0)}\right)\right)$$

for $q_i < n$ or

$$P_i = \left(0, \vartheta, \left|\frac{1}{\mu_i(0)}\right|, arg\left(\frac{1}{\mu_i(0)}\right)\right)$$

for $q_i = n$ respectively. By plugging P_i in $(g \circ e)(u_{q_i+1}, u_{q_i+1}) = \frac{1}{u_{q_i+1}^{q_i} v_{q_i+1}^{q_i+1}}$ for $q_i < n$

(respectively $(g \circ e)(u_{\tilde{n}}, v_{\tilde{n}}) = \frac{1}{u_{\tilde{n}}^n v_{\tilde{n}}}$ for $q_i = n$) we get:

$$\begin{aligned} P_i \in B_0^\vartheta &\Leftrightarrow e^{(g\circ e)(P_i)} \in \mathcal{A}^{mod\ D'_Z} \\ &\Leftrightarrow -q_i\vartheta - arg\left(\frac{1}{\mu_i(0)}\right) \in \left(\frac{\pi}{2}, \frac{3\pi}{2}\right) \\ &\Leftrightarrow arg(\mu_i(0)) - q_i\vartheta \in \left(\frac{\pi}{2}, \frac{3\pi}{2}\right) \end{aligned}$$

On the other hand by Remark 3.6 we know (since $q_i > 0$):

$$\psi_i \leq_\vartheta 0 \Leftrightarrow arg(\mu_i(0)) - q_i\vartheta \in \left(\frac{\pi}{2}, \frac{3\pi}{2}\right)$$

and we conclude

$$P_i \in B_0^\vartheta \Leftrightarrow \psi_i \leq_\vartheta 0.$$

For $\psi \neq 0$ and $\psi \neq \psi_i$ there are different cases to consider. By plugging in

$$P_i = \left(0, \theta_{u_{q_i+1}}, \left|\frac{1}{\mu_i(0)}\right|, arg\left(\frac{1}{\mu_i(0)}\right)\right)$$

in the proper $g \circ e$ we get:

- $q_i < q$:

$$P_i \in B_\psi^\vartheta \Leftrightarrow arg(-\mu(0)) - q\vartheta \in \left(\frac{\pi}{2}, \frac{3\pi}{2}\right)$$

- $q_i = q$:

$$\begin{aligned} P_i \in B_\psi^\vartheta &\Leftrightarrow arg\left(1 - \frac{\mu(0)}{\mu_i(0)}\right) - q\vartheta - arg\left(\frac{1}{\mu_i(0)}\right) \in \left(\frac{\pi}{2}, \frac{3\pi}{2}\right) \\ &\Leftrightarrow arg(\mu_i(0) - \mu(0)) - q\vartheta \in \left(\frac{\pi}{2}, \frac{3\pi}{2}\right) \end{aligned}$$

- $q_i > q$:

$$P_i \in B_\psi^\vartheta \Leftrightarrow -q_i\vartheta - arg\left(\frac{1}{\mu_i(0)}\right) \in \left(\frac{\pi}{2}, \frac{3\pi}{2}\right) \Leftrightarrow arg(\mu_i(0)) - q_i\vartheta \in \left(\frac{\pi}{2}, \frac{3\pi}{2}\right)$$

Referring to Remark 3.6 this verifies the claim for $\psi_i \neq \psi$.
Finally we have to consider $\psi_i = \psi$. But this is obvious: $\psi_i \leq_\vartheta \psi$ for all ϑ. On the other hand $P = \left(0, \theta_{u'_{\tilde{q}}}, 0\right) \in B_\psi^\vartheta$, because $(g \circ e)\left(\tilde{u}'_q, \tilde{v}'_q\right)$ is holomorphic for all ϑ.

□

Lemma 3.19: *Let $\widetilde{P}_j$ denote the intersection points of $\widetilde{S}_j$ with $\left(\widetilde{p \circ e}\right)^{-1}(\vartheta)$. Then we have:*

$$\left\{\widetilde{P}_j \mid j = 1, \dots, J\right\} \subset B_\psi^\vartheta \Leftrightarrow 0 \leq_\vartheta \psi$$

Proof:

- $\psi = 0$: We have $g(t, y) = \frac{1}{y}$. $(g \circ e)(t, y) = \frac{1}{v_1}$ is holomorphic in every point $(0, \theta_{u_1}, |v_1|, \theta_{v_1})$ where $|v_1| \neq 0$. This implies $\left\{\widetilde{P}_j\right\} \subset B_0^\vartheta$ for all ϑ. On the other hand $0 \leq_\vartheta 0$ for all ϑ.
- $\psi \neq 0$: $(g \circ c)(t, y) = \frac{u_1^q v_1^{q-1} - \mu(u_1 v_1)}{u_1^q v_1^q}$ and

$$(0, \theta_{u_1}, |v_1|, \theta_{v_1}) \in B_\psi^\vartheta \Leftrightarrow arg(-\mu(0)) - q\vartheta \in \left(\frac{\pi}{2}, \frac{3\pi}{2}\right)$$

  On the other hand

$$0 \leq_\vartheta \psi \Leftrightarrow arg(-\mu(0)) - q\vartheta \in \left(\frac{\pi}{2}, \frac{3\pi}{2}\right)$$

  Thus we can conclude

$$\{\widetilde{P}_j\} \subset B_\psi^\vartheta \Leftrightarrow arg(-\mu(0)) - q\vartheta \in \left(\frac{\pi}{2}, \frac{3\pi}{2}\right) \Leftrightarrow 0 \leq_\vartheta \psi$$

□

Definition 3.20: We define the following sets:

- $\mathcal{P}^\vartheta := \{\widetilde{P}_j \mid j = 1 \dots, J\} \cup \{P_i \mid i = 1, \dots I\} \subset (p \circ e)^{-1}(\vartheta)$
- $\mathcal{P}_\psi^\vartheta := \mathcal{P}^\vartheta \cap B_\psi^\vartheta$

Corollary 3.21: *By the previous lemma we can conclude*

- $\psi = 0$:

$$\mathcal{P}_0^\vartheta = \{\widetilde{P}_j \mid j = 1 \dots, J\} \cup \{P_i \mid P_i \in B_0^\vartheta\}$$

- $\psi \neq 0$, $\vartheta \in \left(\frac{\frac{\pi}{2} + arg(-\mu(0))}{q}, \frac{\frac{3\pi}{2} + arg(-\mu(0))}{q}\right) \; mod \; \frac{2\pi}{q}$:

$$\mathcal{P}_\psi^\vartheta = \{\widetilde{P}_j \mid j = 1 \dots, J\} \cup \{P_i \mid P_i \in B_\psi^\vartheta\}$$

- $\psi \neq 0$, $\vartheta \in \left(\frac{-\frac{\pi}{2}+arg(-\mu(0))}{q}, \frac{\frac{\pi}{2}+arg(-\mu(0))}{q}\right)$ *mod* $\frac{2\pi}{q}$:

$$\mathcal{P}_\psi^\vartheta = \{P_i \mid P_i \in B_\psi^\vartheta\}$$

Proof: Follows directly from the proof of Lemma 3.19.

□

Lemma 3.22: *Let* $\psi = 0$. *Then the fundamental group* $\pi_1\left(B_0^\vartheta \setminus \mathcal{P}_0^\vartheta\right)$ *is a free group of rank* $\#\left(\mathcal{P}_0^\vartheta\right)$.

Proof: This follows from the fact, that B_0^ϑ emerges from glueing the following sets of points:

- $M_1 = \{\zeta = (0, \theta_{u_1}, |v_1|, \theta_{v_1}) \mid v_1 \neq 0\}$
- $M_2 = \{\zeta = (0, \theta_{u_1}, 0, \theta_{v_1}) \mid \theta_{v_1} \in \left(\frac{\pi}{2}, \frac{3\pi}{2}\right)\}$
- $M_3 = \{\zeta = (0, \theta_{u_k}, |v_k|, \theta_{v_k}) \mid \theta_{v_k} \in \left(\frac{\pi}{2}, \frac{3\pi}{2}\right)\}$
- $M_4 = \{\zeta = (0, \theta_{\tilde{u}_n}, |\tilde{v}_n|, \theta_{\tilde{v}_n}) \mid \theta_{\tilde{v}_n} \in \left(\frac{\pi}{2} - n\vartheta, \frac{3\pi}{2} - n\vartheta\right)\}$

Since M_1 contains the $\widetilde{P}_j$'s and M_2, M_3, M_4 are simply connected and contain the relevant P_i's, this shows the claim.

□

Lemma 3.23: *Let* $\psi \neq 0$, *i. e.* ψ *is given by* $\psi(t) = \mu(t)\, t^{-q}$. *Then* $\pi_1\left(B_\psi^\vartheta \setminus \mathcal{P}_\psi^\vartheta\right)$ *is a free group of rank* $\#\left(\mathcal{P}_\psi^\vartheta\right)$.

Proof: Explicitely we have to prove:

1. For $\vartheta \in \left(\frac{\frac{\pi}{2}+arg(-\mu(0))}{q}, \frac{\frac{3\pi}{2}+arg(-\mu(0))}{q}\right)$ *mod* $\frac{2\pi}{q}$:

$$\pi_1\left(B_\psi^\vartheta \setminus \mathcal{P}_\psi^\vartheta\right) = \#\{\widetilde{P}_j\} + \#\{P_i \mid P_i \in B_\psi^\vartheta\}$$

2. For $\vartheta \notin \left(\frac{\frac{\pi}{2}+arg(-\mu(0))}{q}, \frac{\frac{3\pi}{2}+arg(-\mu(0))}{q}\right)$ $mod\ \frac{2\pi}{q}$:

$$\pi_1\left(B_\psi^\vartheta \setminus \mathcal{P}_\psi^\vartheta\right) = \#\{P_i \mid P_i \in B_\psi^\vartheta\}$$

For $k \leq q$ we have:

$$\zeta = (0, \theta_{u_k}, |v_k|, \theta_{v_k}) \in B_\psi^\vartheta \Leftrightarrow arg\left(-\mu\left(0\right)\right) - q\vartheta \in \left(\frac{\pi}{2}, \frac{3\pi}{2}\right)$$

This explains the sub-division into the two cases above.
Let $q < n$. Then B_ψ^ϑ consists of the following sets:

- $k \leq q$: $M_1 = \{\zeta = (0, \theta_{u_k}, |v_k|, \theta_{v_k})\}$ if $\vartheta \in \left(\frac{\frac{\pi}{2}+arg(-\mu(0))}{q}, \frac{\frac{3\pi}{2}+arg(-\mu(0))}{q}\right)$ $mod\ \frac{2\pi}{q}$ and $M_1 = \emptyset$ otherwise
- $k = q+1$: $M_2 = \left\{\zeta = (0, \theta_{u_k}, |v_k|, \theta_{v_k}) \mid arg\left(1 - \mu\left(0\right) v_k\right) - q\vartheta - \theta_{v_k} \in \left(\frac{\pi}{2}, \frac{3\pi}{2}\right)\right\}$
- $k > q+1$: $M_3 = \left\{\zeta = (0, \theta_{u_k}, |v_k|, \theta_{v_k}) \mid \theta_{v_k} \in \left(\frac{\pi}{2} - (k-1)\vartheta, \frac{3\pi}{2} - (k-1)\vartheta\right)\right\}$
- $k = n$: $M_4 = \left\{\zeta = (0, \vartheta, |\tilde{v}_n|, \theta_{\tilde{v}_n}) \mid \theta_{\tilde{v}_n} \in \left(\frac{\pi}{2} - n\vartheta, \frac{3\pi}{2} - n\vartheta\right)\right\}$
- Branching in $P = \left(0, \theta_{u_k}, \left|\frac{1}{\mu(0)}\right|, arg\left(\frac{1}{\mu(0)}\right)\right)$, $k = q+1$
  - $M_1' = \left\{\zeta = (0, \theta_{u_s'}, |v_s'|, \theta_{v_s'}) \mid \pi + 2\,arg\left(\mu\left(0\right)\right) - (q-s)\,\theta_{u_k} - \theta_{u_s'} \in \left(\frac{\pi}{2}, \frac{3\pi}{2}\right)\right\}$
    Remark that θ_{u_k} is fixed by the given branching point P: $\theta_{u_k} = \vartheta - arg\left(\frac{1}{\mu(0)}\right)$
  - $M_2' = \left\{\zeta = \left(0, \theta_{\tilde{u}_q'}, \left|\tilde{v}_q'\right|, \theta_{\tilde{v}_q'}\right)\right\}$

We see that B_ψ^ϑ contains all the $\widetilde{P}_j$s depending on ϑ. Furthermore, obviously M_3, M_4 and M_1', M_2' are simply connected. So we only have to take care about M_2. We also can read M_2 in coordinates (u_q, v_q):

$$M_2 = \left\{\left(|u_q|, \theta_{u_q}, 0, \theta_{v_q}\right) \mid arg\left(u_q - \mu\left(0\right)\right) - q\vartheta \in \left(\frac{\pi}{2}, \frac{3\pi}{2}\right)\right\}$$

But this defines a half-plane and therefore is simply connected.
It is obvious that M_1, M_2, M_3, M_4 and M_1', M_2' are glued sets. Therefore it remains to show that M_2 M_1' glue. Because of the coordinate change ($u_k = u_1'v_1', v_k - \frac{1}{\mu(u_k v_k)}$) we can rewrite $\zeta = (0, \theta_{u_k}, |v_k|, \theta_{v_k}) \in M_2$ to $\left(0, \theta_{u_1'} + \theta_{v_1'}, \left|v_1' + \frac{1}{\mu(0)}\right|, arg\left(v_1' + \frac{1}{\mu(0)}\right)\right)$ with

the following condition:

$$arg\left(1-\mu(0)\left(v_1'+\frac{1}{\mu(0)}\right)\right)-q\theta_{u_1'}-q\theta_{v_1'}-arg\left(v_1'+\frac{1}{\mu(0)}\right)\in\left(\frac{\pi}{2},\frac{3\pi}{2}\right)$$
$$\Leftrightarrow arg\left(\frac{-\mu(0)\,v_1'}{v_1'+\frac{1}{\mu(0)}}\right)-q\theta_{u_1'}-(q-1)\,\theta_{v_1'}-\theta_{v_1'}\in\left(\frac{\pi}{2},\frac{3\pi}{2}\right)$$
$$\Leftrightarrow arg\left(\frac{-\mu(0)}{v_1'+\frac{1}{\mu(0)}}\right)-q\theta_{u_1'}-(q-1)\,\theta_{v_1'}\in\left(\frac{\pi}{2},\frac{3\pi}{2}\right)$$
$$\Leftrightarrow \pi+arg\left(\mu(0)\right)+arg\left(\frac{1}{v_1'+\frac{1}{\mu(0)}}\right)-(q-1)\,\theta_{u_k}-\theta_{u_1'}\in\left(\frac{\pi}{2},\frac{3\pi}{2}\right)$$

But this is exactly the condition for $\left(0,\theta_{u_1'},|v_1'|=0,\theta_{v_1'}\right)\in M_1'$. Thus these two sets glue.

The proof for $q=n$ works similarly, but simplifies to:

- $k\leq n$: $M_1=\{\zeta=(0,\theta_{u_k},|v_k|,\theta_{v_k})\}$ if $\vartheta\in\left(\frac{\frac{\pi}{2}+arg(-\mu(0))}{n},\frac{\frac{3\pi}{2}+arg(-\mu(0))}{n}\right)\ mod\ \frac{2\pi}{n}$ and $M_1=\emptyset$ otherwise
- $k=n$: $M_2=\left\{\zeta=(0,\vartheta,|v_{\tilde{n}}|,\theta_{v_{\tilde{n}}})\mid arg\left(1-\mu(0)\,v_{\tilde{n}}\right)-n\vartheta-\theta_{v_{\tilde{n}}}\in\left(\frac{\pi}{2},\frac{3\pi}{2}\right)\right\}$
- Branching in $P=\left(0,\vartheta,\left|\frac{1}{\mu(0)}\right|,arg\left(\frac{1}{\mu(0)}\right)\right)$
  - $M_1'=\left\{\zeta=(0,\theta_{u_s'},|v_s'|,\theta_{v_s'})\mid \pi+2\,arg\left(\mu(0)\right)-(n-s)\,\vartheta-\theta_{u_s'}\in\left(\frac{\pi}{2},\frac{3\pi}{2}\right)\right\}$
  - $M_2'=\{\zeta=(0,\vartheta,|\tilde{v}_n'|,\theta_{\tilde{v}_n'})\}$

□

Descriptively, this means, that there are no other 'holes' in B_ψ^ϑ than the singularities P_i, which arise from the intersection with S_i and possibly – depending on the choice of ϑ if $\psi\neq 0$ – the singularities $\widetilde{P}_j$, which arise from the intersection with $\widetilde{S}_j$. Thus we can interpret B_ψ^ϑ as one of the following topological spaces:

1. $\psi=0$

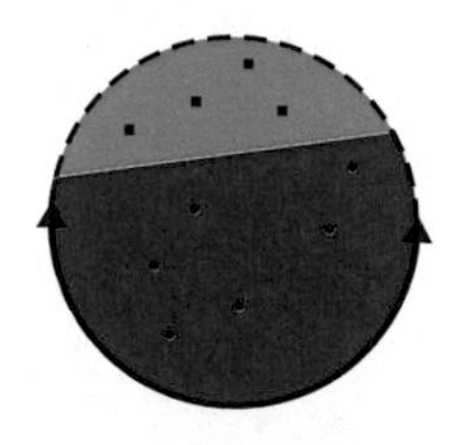

2. $\psi \neq 0$, $\vartheta \in \left(\frac{\frac{\pi}{2}+arg(-\mu(0))}{q}, \frac{\frac{3\pi}{2}+arg(-\mu(0))}{q}\right) \; mod \; \frac{2\pi}{q}$

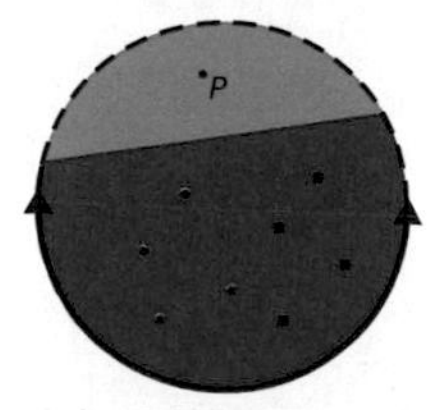

3. $\psi \neq 0$, $\vartheta \in \left(\frac{\frac{-\pi}{2}+arg(-\mu(0))}{q}, \frac{\frac{\pi}{2}+arg(-\mu(0))}{q}\right) \; mod \; \frac{2\pi}{q}$

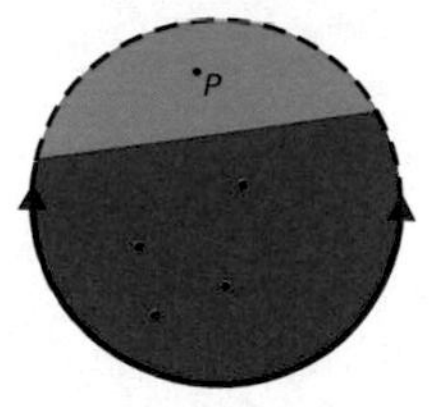

The green part represents the points of B_{ψ}^{ϑ}, where $(e \circ g)$ is holomorphic, the red part stands for the points, where $(e \circ g)$ is good. Furthermore the black dots show the points P_i, in contrast the points $\ddot{P}_j$ are symbolized by the black squares.
Remark that an open interval of the boundary of B_{ψ}^{ϑ} lies in B_{ψ}^{ϑ}! This holds because for $k = n$ we have

$$(0, \vartheta, 0, \theta_{\tilde{v}_n}) \in B_{\psi}^{\vartheta} \Leftrightarrow \theta_{\tilde{v}_n} \in \left(\frac{\pi}{2} - n\vartheta, \frac{3\pi}{2} - n\vartheta\right)$$

both if $q < n$ and if $q = n$.

3.3.4 Dimension of $\mathbb{H}^1_c\left(B^{\vartheta}_{\psi}, \mathcal{F}_{\psi}\right)$

Proposition 3.24: *On $B^{\vartheta}_{\psi} \setminus \mathcal{P}^{\vartheta}_{\psi}$ the perverse sheaf $\mathcal{F}_{\psi}$ has cohomology in degree 0 at most.*

Proof: We know that $e^{+}\mathcal{M}$ has regular singularities along the divisor and that $\mathcal{E}^{g(t,y)}$ is good or even holomorphic. Moreover the divisor is normal crossing except possibly at the intersection points P_i, $\widetilde{P}_j$. Thus we can apply [Sab13], Corollary 12.7.

□

Lemma 3.25: *Let Δ be an open disc round the singular point 0 and Δ' an open subset of the closure $\overline{\Delta}$, consisting of Δ and a connected open subset of $\partial\overline{\Delta}$. Let $\mathcal{M}$ be a regular singular $\mathcal{D}_{\Delta'}$-module and consider a perverse sheaf $\mathcal{F}$ with singularity only in 0. Then $\mathbb{H}^k_c(\Delta', \mathcal{F}) = 0$ for $k \neq 1$ and* dim $\mathbb{H}^1_c(\Delta', \mathcal{F})$ *is equal to the dimension of the vanishing cycle space at 0.*

Proof: We will consider the following diagram of maps and spaces

$$\Delta' \setminus \{0\} \overset{i}{\hookrightarrow} \Delta' \overset{\beta}{\hookrightarrow} \overline{\Delta} \overset{j}{\hookleftarrow} Z$$

At first, let's remark:

- Let $\mathcal{F}$ be a complex of sheaves on Δ'. Then we have an exact sequence of complexes of sheaves on $\overline{\Delta}$:

$$0 \to \beta_!\mathcal{F} \to \beta_*\mathcal{F} \to j_*j^{-1}\beta_*\mathcal{F} \to 0$$

  and thus an exact sequence of cohomologies:

$$\begin{aligned} \ldots \to \mathbb{H}^n\left(\overline{\Delta}, \beta_!\mathcal{F}\right) \to \mathbb{H}^n\left(\overline{\Delta}, \beta_*\mathcal{F}\right) \to \mathbb{H}^n\left(\overline{\Delta}, j_*j^{-1}\beta_*\mathcal{F}\right) \to \\ \to \mathbb{H}^{n+1}\left(\overline{\Delta}, \beta_!\mathcal{F}\right) \to \mathbb{H}^{n+1}\left(\overline{\Delta}, \beta_*\mathcal{F}\right) \to \mathbb{H}^{n+1}\left(\overline{\Delta}, j_*j^{-1}\beta_*\mathcal{F}\right) \to \ldots \end{aligned}$$

- Since perverse sheaves form an abelian category, each object obtains a Jordan Hölder sequence, i.e. every perverse sheaf is a successive extension of simple perverse sheaves. These have one of the following form

- $0 \to i_*\mathcal{L} \to 0 \to 0$, where $\mathcal{L}$ is a local system on $\Delta' \setminus 0$.
- $0 \to 0 \to i_{0*}F \to 0$, where i_0 denotes the inclusion $i_0 : \{0\} \hookrightarrow \Delta'$ and thus $i_{0*}F$ is a skyscraper sheaf with support in 0.

- If $\mathcal{F}$ (e.g. $i_*\mathcal{L}$ or $i_{0*}F[-1]$) is an ordinary sheaf, we have $\mathbb{H}^\bullet = H^\bullet$.

Thus it is enough to compute dim $H^1_c(\Delta', \mathcal{F})$ for $\mathcal{F} = i_*\mathcal{L}$ and $\mathcal{F} = i_{0*}F[-1]$.

Let $\mathcal{F} = i_*\mathcal{L}$ with $\mathcal{L}$ a local system on $\Delta' \setminus \{0\}$. Obviously $H^{-1}\left(\overline{\Delta}, \beta_!\mathcal{F}\right) = 0$ and the same holds for $\beta_*\mathcal{F}$, $j_*j^{-1}\beta_*\mathcal{F}$. Moreover we know:

- $H^i\left(\overline{\Delta}, \beta_*\mathcal{F}\right) = H^i(\Delta', \mathcal{F}) = 0$ for $i > 1$ because $\mathcal{F}$ is a perverse sheaf and therefore has cohomology in degree $0, 1$ at most.
- $H^i\left(\overline{\Delta}, j_*j^{-1}\beta_*\mathcal{F}\right) = H^i\left(Z, j^{-1}\beta_*\mathcal{F}\right) = 0$ for $i \leq 1$. This holds because Z is contractible and $j^{-1}\beta_*\mathcal{F}$ is a constant sheaf on Z (and therefore flasque).

Thus the cohomology sequence reduces to:

$$0 \to H^0\left(\overline{\Delta}, \beta_!\mathcal{F}\right) \to H^0\left(\overline{\Delta}, \beta_*\mathcal{F}\right) \to H^0\left(\overline{\Delta}, j_*j^{-1}\beta_*\mathcal{F}\right) \to H^1\left(\overline{\Delta}, \beta_!\mathcal{F}\right) \to H^1\left(\overline{\Delta}, \beta_*\mathcal{F}\right) \to 0 \tag{3.2}$$

We will compute the cohomology groups explicitely:

1. $H^0\left(\overline{\Delta}, \beta_!\mathcal{F}\right) = 0$:
   $H^0\left(\overline{\Delta}, \beta_!\mathcal{F}\right) = \beta_!\mathcal{F}\left(\overline{\Delta}\right)$ and $\beta_!\mathcal{F}$ is given by:

$$\beta_!\mathcal{F}(U) = \begin{cases} \mathcal{F}(U) & \text{for } U \cap Z = \emptyset \\ 0 & \text{for } U \cap Z \neq \emptyset \end{cases}$$

   Furthermore we have $\overline{\Delta} \cap Z \neq \emptyset$.

2. $H^0\left(\overline{\Delta}, \beta_*\mathcal{F}\right) = ker\,(T-1)$ (where T denotes the monodromy automorphism corresponding to the local system $\mathcal{L}$):

$$H^0\left(\overline{\Delta}, \beta_*\mathcal{F}\right) = H^0(\Delta', \mathcal{F}) \overset{\mathcal{F}=i_*\mathcal{L}}{=} H^0(\Delta' \setminus \{0\}, \mathcal{L}) = ker\,(T-1)$$

3. $H^0\left(\overline{\Delta}, j_*j^{-1}\beta_*\mathcal{F}\right) = \mathcal{L}_{x_0}$ for some $x_0 \in U \cap \Delta' \setminus \{0\}$:

$$\begin{aligned} H^0\left(\overline{\Delta}, j_*j^{-1}\beta_*\mathcal{F}\right) &= H^0\left(Z, j^{-1}\beta_*\mathcal{F}\right) = j^{-1}\beta_*\mathcal{F}(Z) \\ &= \varinjlim_{Z\subset U} \beta_*\mathcal{F}(U) = \varinjlim_{Z\subset U} \mathcal{F}(U\cap\Delta') = \varinjlim_{Z\subset U} \mathcal{L}(U\cap\Delta'\setminus\{0\}) \\ &= \varinjlim_{Z\subset U \text{ small enough}} \mathcal{L}_{x_0} \\ &= \mathcal{L}_{x_0} \end{aligned}$$

4. $H^1\left(\overline{\Delta}, \beta_*\mathcal{F}\right) = H^1\left(\overline{\Delta}, (\beta\circ i)_*\mathcal{L}\right)$:
By considering the Leray spectral sequence, we get

$$E_2^{p,q} = H^p\left(\overline{\Delta}, R^q(\beta\circ i)_*\mathcal{L}\right) \Rightarrow E^{p+q} = H^{p+q}(\Delta'\setminus\{0\}, \mathcal{L})$$

and the 5-term exact sequence

$$0 \to E_2^{1,0} \to E^1 \to E_2^{0,1} \to E_2^{2,0} \to E^2$$

that is

$$\begin{aligned} 0 &\to H^1\left(\overline{\Delta}, (\beta\circ i)_*\mathcal{L}\right) \to H^1(\Delta'\setminus\{0\}, \mathcal{L}) \to H^0\left(\overline{\Delta}, R^1(\beta\circ i)_*\mathcal{L}\right) \to \\ &\to H^2\left(\overline{\Delta}, (\beta\circ i)_*\mathcal{L}\right) \to H^2(\Delta'\setminus\{0\}, \mathcal{L}) \to 0 \end{aligned} \tag{3.3}$$

We have

- $H^1(\Delta'\setminus\{0\}, \mathcal{L}) = coker\,(T-1)$
- $H^2\left(\overline{\Delta}, (\beta\circ i)_*\mathcal{L}\right) = 0$
- $H^2(\Delta'\setminus\{0\}, \mathcal{L}) = 0$

It remains to compute $H^0\left(\overline{\Delta}, R^1(\beta\circ i)_*\mathcal{L}\right)$.
$R^1(\beta\circ i)_*\mathcal{L}$ can be identified as the sheafification of the cohomology presheaf $\mathcal{H}^1$ (cf. [Ive86], p. 105).

$$R^1(\beta\circ i)_*\mathcal{L} : \left(U \mapsto H^1\left((\beta\circ i)^{-1}(U), \mathcal{L}\right)\right)^\sim = H^1(U\cap\Delta'\setminus\{0\}, \mathcal{L})^\sim$$

We get $(R^1(\beta\circ i)_*\mathcal{L})_0 = coker\,(T-1)$ and we have an isomorphism on the stalk

$$R^1(\beta\circ i)_*\mathcal{L}\left(\overline{\Delta}\right) \xrightarrow{\cong} \left(R^1(\beta\circ i)_*\mathcal{L}\right)_0$$

Thus we can conclude $R^1 (\beta \circ i)_* \mathcal{L}\left(\overline{\Delta}\right) \cong coker\,(T-1)$ By comparing dimensions it follows from the sequence (3.3)

$$H^1\left(\overline{\Delta}, \beta_* \mathcal{F}\right) = H^1\left(\Delta', \mathcal{F}\right) = 0$$

Now we can compute the dimension of $H^1_c\left(\Delta', \mathcal{F}\right) = H^1\left(\overline{\Delta}, \beta_! \mathcal{F}\right)$ via the sequence (3.2):

$$\begin{aligned} \dim\, H^1\left(\overline{\Delta}, \beta_! \mathcal{F}\right) &= -\dim\, H^0\left(\overline{\Delta}, \beta_* \mathcal{F}\right) + \dim\, H^0\left(\overline{\Delta}, j_* j^{-1} \beta_* \mathcal{F}\right) \\ &= -\dim\left(ker\,(T-1)\right) + \dim\, \mathcal{L}_{x_0} \\ &= \dim\left(im\,(T-1)\right) \end{aligned}$$

Finally, we know that $\dim\left(im\,(T-1)\right)$ is equal to $\dim\, \Phi_0$, the dimension of the vanishing cycle space in the singularity 0.
Let now $\mathcal{F} : 0 \to 0 \to i_{0*} F \to 0$. Then we have $\mathbb{H}^{i+1}\left(\overline{\Delta}, \beta_! \mathcal{F}\right) = H^i\left(\overline{\Delta}, \beta_! i_{0*} F\right)$. Obviously $H^i\left(\overline{\Delta}, \beta_! i_{0*} F\right) = 0$ for $i \neq 0$ and $H^0\left(\overline{\Delta}, \beta_! i_{0*} F\right) = F = \Phi_0$

□

Using the Lemma above, we can state the following theorem about the dimension of $H^1_c\left(B^\vartheta_\psi, \mathcal{F}^\bullet_\psi\right)$:

Theorem 3.26: *We have the following results for the dimension of* $H^1_c\left(B^\vartheta_\psi, \mathcal{F}^\bullet_\psi\right)$*:*

1. $\psi = 0$*:*

$$\dim\, \mathbb{H}^1_c\left(B^\vartheta_0, \mathcal{F}^\bullet_0\right) = \sum_{P \in \mathcal{P}^\vartheta_0} \dim\, \Phi_P = \sum_j \dim\, \Phi_{\widetilde{P}_j} + \sum_{\{i | P_i \in B^\vartheta_0\}} \dim\, \Phi_{P_i}$$

2. $\psi \neq 0$*:*

$$\dim\, \mathbb{H}^1_c\left(B^\vartheta_\psi, \mathcal{F}^\bullet_\psi\right) = \sum_{P \in \mathcal{P}^\vartheta_\psi} \dim\, \Phi_P,$$

i. e. explicitely:

- $\vartheta \in \left(\frac{\frac{\pi}{2} + arg(-\mu(0))}{q}, \frac{\frac{3\pi}{2} + arg(-\mu(0))}{q}\right)$ *mod* $\frac{2\pi}{q}$*:*

$$\dim\, \mathbb{H}^1_c\left(B^\vartheta_\psi, \mathcal{F}^\bullet_\psi\right) = \sum_j \dim\, \Phi_{\widetilde{P}_j} + \sum_{\{i | P_i \in B^\vartheta_\psi\}} \dim\, \Phi_{P_i}$$

- $\vartheta \in \left(\frac{-\frac{\pi}{2}+arg(-\mu(0))}{q}, \frac{\frac{\pi}{2}+arg(-\mu(0))}{q}\right) \, mod \, \frac{2\pi}{q}$:

$$\dim \mathbb{H}^1_c\left(B^{\vartheta}_{\psi}, \mathcal{F}^{\bullet}_{\psi}\right) = \sum_{\{i|P_i \in B^{\vartheta}_{\psi}\}} \dim \Phi_{P_i}$$

Proof: For abbreviation we will write B for B^{ϑ}_{ψ} and $\mathcal{F}$ for $\mathcal{F}_{\psi}$.
We can find an open covering (U_1, U_2) of $B = U_1 \cup U_2$ with:

- U_1 is homeomorphic to Δ', i. e. U_1 contains exactly one singular point $P \in \mathcal{P}^{\vartheta}_{\psi}$
- $U_2 \cap \partial B \neq \emptyset$
- $U_1 \cap U_2$ contains no singular point of B.

We get the following diagram:

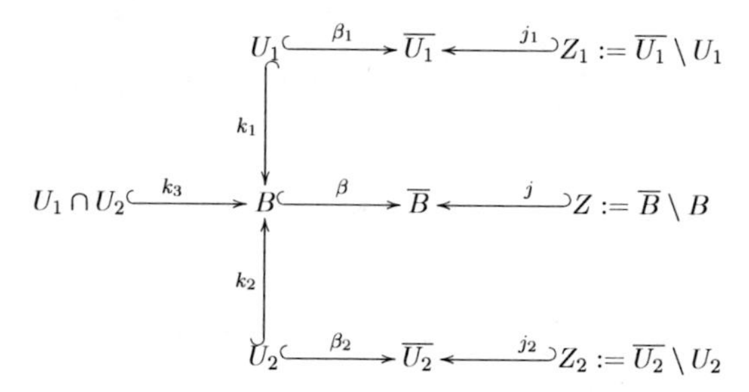

Using Mayer-Vietoris we get an exact sequence of sheaves on $\overline{B}$

$$0 \to \beta_! \mathcal{F} \to \beta_! k_{1*} k_1{}^* \mathcal{F} \oplus \beta_! k_{2*} k_2{}^* \mathcal{F} \to \beta_! k_{3*} k_3{}^* \mathcal{F} \to 0$$

We have $\mathbb{H}^i\left(\overline{B}, \beta_! \bullet\right) = 0$ for $i \neq 0, 1$, since $\bullet$ are perverse sheaves and thus have only $\mathbb{H}^0, \mathbb{H}^1$. Moreover $\mathbb{H}^0$ vanishes because of $\beta_!$. Thus we get an exact sequence

$$0 \to \mathbb{H}^1\left(\overline{B}, \beta_! \mathcal{F}\right) \to \mathbb{H}^1\left(\overline{B}, \beta_! k_{1*} k_1{}^* \mathcal{F}\right) \oplus \mathbb{H}^1\left(\overline{I}, \beta_! k_{2*} k_2{}^* \mathcal{F}\right) \to \mathbb{H}^1\left(\overline{B}, \beta_! k_{3*} k_3{}^* \mathcal{F}\right) \to 0$$

Hence have the following formula for the dimensions:

$$\dim \mathbb{H}^1\left(\overline{B}, \beta_! \mathcal{F}\right) = \dim \mathbb{H}^1\left(\overline{B}, \beta_! k_{1*} k_1{}^* \mathcal{F}\right) + \dim \mathbb{H}^1\left(\overline{B}, \beta_! k_{2*} k_2{}^* \mathcal{F}\right) - \dim \mathbb{H}^1\left(\overline{B}, \beta_! k_{3*} k_3{}^* \mathcal{F}\right)$$

Claim:

1. $\dim\ \mathbb{H}^1\left(\overline{B}, \beta_! k_{1*} k_1{}^* \mathcal{F}\right) = \dim\ \mathbb{H}^1\left(\overline{\Delta}, \beta_{1!} k_1{}^* \mathcal{F}\right)$
2. $\dim\ \mathbb{H}^1\left(\overline{B}, \beta_! k_{2*} k_2{}^* \mathcal{F}\right) = \dim\ \mathbb{H}^1\left(\overline{U_2}, \beta_{2!} k_2{}^* \mathcal{F}\right)$
3. $\mathbb{H}^1\left(\overline{B}, \beta_! k_{3*} k_3{}^* \mathcal{F}\right) = 0$

We will prove (1.) ((2.) works similarly):
We have the following exact sequences $(\star)$

$$0 \to \mathbb{H}^0\left(\overline{B}, \beta_* k_{1*} k_1{}^* \mathcal{F}\right) \to \mathbb{H}^0\left(\overline{B}, j_* j^* \beta_* k_{1*} k_1{}^* \mathcal{F}\right) \to \mathbb{H}^1\left(\overline{B}, \beta_! k_{1*} k_1{}^* \mathcal{F}\right) \to \mathbb{H}^1\left(\overline{B}, \beta_* k_{1*} k_1{}^* \mathcal{F}\right) \to 0$$

and

$$0 \to \mathbb{H}^0\left(\overline{U_1}, \beta_{1*} k_1{}^* \mathcal{F}\right) \to \mathbb{H}^0\left(\overline{U_1}, j_{1*} j_1{}^* \beta_{1*} k_1{}^* \mathcal{F}\right) \to \mathbb{H}^1\left(\overline{U_1}, \beta_{1!} k_1{}^* \mathcal{F}\right) \to \mathbb{H}^1\left(\overline{U_1}, \beta_{1*} k_1{}^* \mathcal{F}\right) \to 0$$

Thus we can deduce:

- $\dim\ \mathbb{H}^0\left(\overline{B}, \beta_* k_{1*} k_1{}^* \mathcal{F}\right) = \dim\ \mathbb{H}^0\left(\overline{U_1}, \beta_{1*} k_1{}^* \mathcal{F}\right) = \dim\ \mathbb{H}^0\left(U_1, k_1{}^* \mathcal{F}\right)$
- $\dim\ \mathbb{H}^1\left(\overline{B}, \beta_* k_{1*} k_1{}^* \mathcal{F}\right) = \dim\ \mathbb{H}^1\left(\overline{U_1}, \beta_{1*} k_1{}^* \mathcal{F}\right) = \dim\ \mathbb{H}^1\left(U_1, k_1{}^* \mathcal{F}\right)$

Furthermore we have

$$\mathbb{H}^0\left(\overline{B}, j_* j^* \beta_* k_{1*} k_1{}^* \mathcal{F}\right) = \varinjlim_{Z \subset U} k_1{}^* \mathcal{F}\left(U \cap U_1\right) = \mathcal{F}_{x_0}$$

just as

$$\mathbb{H}^0\left(\overline{U_1}, j_{1*} j_1{}^* \beta_{1*} k_1{}^* \mathcal{F}\right) = \varinjlim_{Z_1 \subset U} k_1{}^* \mathcal{F}\left(U \cap U_1\right) = \mathcal{F}_{x_0}$$

Hence we can conclude:

$$\dim\ \mathbb{H}^0\left(\overline{B}, j_* j^* \beta_* k_{1*} k_1{}^* \mathcal{F}\right) = \dim\ \mathbb{H}^0\left(\overline{U_1}, j_{1*} j_1{}^* \beta_{1*} k_1{}^* \mathcal{F}\right)$$

By computing the dimensions via the above exact sequences $(\star)$ we get the following result:

$$\dim\ \mathbb{H}^1\left(\overline{B}, \beta_! k_{1*} k_1{}^* \mathcal{F}\right) = \dim\ \mathbb{H}^1\left(\overline{U_1}, \beta_{1!} k_1{}^* \mathcal{F}\right)$$

This proofs Claim 1.
Proof for Claim 3:
$U_1 \cap U_2$ doesn't contain any singular points, i.e. $k_3{}^* \mathcal{F}$ is a local system $\mathcal{L}$. Thus we have:

$$\begin{array}{ccccccccc} 0 \to \mathbb{H}^0\left(\overline{B}, \beta_* k_{3*} k_3{}^* \mathcal{F}\right) & \to & \mathbb{H}^0\left(\overline{B}, j_* j^* \beta_* k_{3*} k_3{}^* \mathcal{F}\right) & \to & \mathbb{H}^1\left(\overline{B}, \beta_! k_{3*} k_3{}^* \mathcal{F}\right) & \to & \mathbb{H}^1\left(\overline{B}, \beta_* k_{3*} k_3{}^* \mathcal{F}\right) \to 0 \\ \| & & \| & & & & \| \\ \mathbb{H}^0\left(U_1 \cap U_2, k_3{}^* \mathcal{F}\right) & & \mathcal{L}_{x_0} & & & & \mathbb{H}^1\left(U_1 \cap U_2, k_3{}^* \mathcal{F}\right) \\ \| & & & & & & \| \\ \mathcal{L}_{x_0} & & & & & & 0 \end{array}$$

This proofs Claim 3: $\mathbb{H}^1\left(\overline{B}, \beta_! k_{3*} k_3{}^* \mathcal{F}\right) = 0$.

From Lemma 3.25 we can deduce:

$$\dim\, \mathbb{H}^1\left(\overline{B}, \beta_! k_{1*} k_2{}^* \mathcal{F}\right) = \dim\, \mathbb{H}^1\left(\overline{\Delta}, \beta_{1!} k_1{}^* \mathcal{F}\right) = \Phi_P$$

By successively splitting U_2 in open sets homeomorphic to Δ' and a repeated use of the above Mayer-Vietoris argument, we can compute

$$\dim\, \mathbb{H}^1\left(\overline{B}, \beta_! k_{2*} k_2{}^* \mathcal{F}\right) = \dim\, \mathbb{H}^1\left(\overline{U_2}, \beta_{2!} k_2{}^* \mathcal{F}\right)$$

via Lemma 3.25 and we get:

$$\dim\, \mathbb{H}^1\left(\overline{B}, \beta_! \mathcal{F}\right) = \sum_{\{P \in \mathcal{P}_I\}} \dim\, \Phi_P$$

Lemma 3.22 and Lemma 3.23 show, which singular points are contained in the relevant $\mathcal{P}_\psi^\vartheta$.

□

Finally, we can show the equality of the dimensions of $(\mathcal{L}_{\leq\psi})_\vartheta$ and $\left(\widetilde{\mathcal{L}}_{\leq\psi}\right)_\vartheta$ for $\vartheta \in \mathbb{S}^1$.

Theorem 3.27: *For $\vartheta \in \mathbb{S}^1$ we have* $\dim\left(\mathcal{L}_{\leq\psi}\right)_\vartheta = \dim\left(\widetilde{\mathcal{L}}_{\leq\psi}\right)_\vartheta$.

Proof: This follows from Corollary 3.7 and Proposition 3.26. Explicitely:

1. $\psi = 0$:

$$\begin{aligned} \dim\left(\widetilde{\mathcal{L}}_{\leq 0}\right)_\vartheta &= \dim\, H_c^1\left(B_0^\vartheta, \mathcal{F}_0\right) \\ &= \sum_j \dim\, \Phi_{\widetilde{P}_j} + \sum_{\{i \mid P_i \in B_0^\vartheta\}} \dim\, \Phi_{P_i} \\ &= \dim\left(\mathcal{L}_{\leq 0}\right)_\vartheta \end{aligned}$$

2. $\psi \neq 0$, $\vartheta \in \left(\frac{\frac{\pi}{2}+arg(-\mu(0))}{q}, \frac{\frac{3\pi}{2}+arg(-\mu(0))}{q}\right) \; mod \; \frac{2\pi}{q}$:

$$\begin{aligned}\dim\left(\widetilde{\mathcal{L}}_{\leq\psi}\right)_{\vartheta} &= \dim\, H^1_c\left(B^{\vartheta}_0, \mathcal{F}_{\psi}\right)\\ &= \sum_j \dim\, \Phi_{\widetilde{P}_j} + \sum_{\{i|P_i\in B^{\vartheta}_{\psi}\}} \dim\, \Phi_{P_i}\\ &= \dim\left(\mathcal{L}_{\leq\psi}\right)_{\vartheta}\end{aligned}$$

3. $\psi \neq 0$, $\vartheta \in \left(\frac{-\frac{\pi}{2}+arg(-\mu(0))}{q}, \frac{\frac{\pi}{2}+arg(-\mu(0))}{q}\right) \; mod \; \frac{2\pi}{q}$:

$$\begin{aligned}\dim\left(\widetilde{\mathcal{L}}_{\leq\psi}\right)_{\vartheta} &= \dim\, H^1_c\left(B^{\vartheta}_{\psi}, \mathcal{F}_{\psi}\right)\\ &= \sum_{\{i|P_i\in B^{\vartheta}_{\psi}\}} \dim\, \Phi_{P_i}\\ &= \dim\left(\mathcal{L}_{\leq\psi}\right)_{\vartheta}\end{aligned}$$

□

Remark 3.28: In an analogous way we can determine $\mathcal{L}$ on $\mathbb{S}_1$. First let us recall that $\mathcal{L} = j_*\mathcal{L}'$, whereby $\mathcal{L}'$ denotes the local system associated to $\mathcal{N} := \mathcal{H}^0 p_+\left(\mathcal{M} \otimes \mathcal{E}^{\frac{1}{y}}\right)$. Consider a small circle $\mathbb{S}^1_{\epsilon}$ around $0 \in \Delta$. Then we have $\mathcal{L}_{|S^1} \cong \mathcal{L}'_{|S^1_{\epsilon}}$. On $\mathbb{S}^1_{\epsilon}$ we can identify $\mathcal{L}' = \mathcal{H}^0\, \mathrm{DR}^{mod\,0}_{\widetilde{\Delta}}(\mathcal{N})$ (since the growing condition $mod\,0$ is irrelevant outside of $\partial\widetilde{\Delta}$). Furthermore in this situation, the isomorphism

$$\Omega : \mathcal{H}^0\, \mathrm{DR}^{mod\,0}_{\widetilde{\Delta}}(\mathcal{N})_{|\mathbb{S}^1_{\epsilon}} \xrightarrow{\cong} \mathcal{H}^1 R\widetilde{p}_*\, \mathrm{DR}^{mod\,D}\left(\mathcal{M} \otimes \mathcal{E}^{\frac{1}{y}}\right)_{|S^1_{\epsilon}}$$

holds (analogously to Theorem 3.8). Now as before we can describe the right hand side in topological terms. Since $\mathcal{E}^{\frac{1}{y}}$ is a good (or even holomorphic) connection, we know that $\mathrm{DR}^{mod\,D}\left(\mathcal{M} \otimes \mathcal{E}^{\frac{1}{y}}\right)$ has cohomology in degree zero at most and thus corresponds to a sheaf. By taking the stalk at a point $\rho \in \mathbb{S}^1_{\epsilon}$ we get

$$\left(\mathcal{H}^0\, \mathrm{DR}^{mod\,0}_{\widetilde{\Delta}}(\mathcal{N})\right)_{\rho} \cong H^1_c\left(\widetilde{p}^{-1}(\rho), \mathrm{DR}^{mod\,D}\left(\mathcal{M} \otimes \mathcal{E}^{\frac{1}{y}}\right)\right).$$

$\widetilde{p}^{-1}(\rho)$ corresponds to the projective line with a real blow-up in the point ∞ (denoted by $\mathbb{S}^1_{\infty}$). We can interpret this space as a disc with boundary $\mathbb{S}^1_{\infty}$. Furthermore by choosing ϵ small enough, we know that every irreducible component S_i resp. $\widetilde{S}_j$ of $SS(\mathcal{M})$ meets

the (inner of the) disc in exactly one point (P_i resp. $\widetilde{P}_j$).
The sheaf is supported everywhere away from the boundary, on the boundary it is supported on an open hemicycle of $\mathbb{S}^1_\infty$ (i.e. the points where it fulfills the moderate growth condition). Thus the dimension of $H^1_c\left(\widetilde{p}^{-1}(\rho), \mathrm{DR}^{mod\,D}\left(\mathcal{M} \otimes \mathcal{E}^{\frac{1}{y}}\right)\right)$ depends on the dimension of the vanishing cycle spaces in the points P_i and $\widetilde{P}_j$:

$$\dim H^1_c\left(\widetilde{p}^{-1}(\rho), \mathrm{DR}^{mod\,D}\left(\mathcal{M} \otimes \mathcal{E}^{\frac{1}{y}}\right)\right) = \dim \sum_{i \in I} \Phi_{P_i} + \sum_{j \in J} \Phi_{\widetilde{P}_j}.$$

Theorem 3.27 proves that the morphism Ω, which is injective according to 3.9, is also surjective and thus an isomorphism. Additionally let us remark that the proof of surjectivity we presented above, provides a topological description of $\widetilde{\mathcal{L}}_{\leq\psi}$ by using the isomorphism

$$\widetilde{\mathcal{L}}_{\leq\psi} \cong H^1_c\left(B^\vartheta_\psi, \mathcal{F}_\psi\right).$$

We will use this topological description in the following chapter and supply it to an explicit example. Moreover, we will see that in this case it even enables us to determine linear Stokes data.

4 Explicit example of the determination of Stokes data

4.1 Stokes-filtered local system

Let again be $X = \Delta \times \mathbb{P}^1$ and $\mathcal{M}$ a meromorphic connection on X of rank r with regular singularities along its divisor. Let the singular locus $SS(\mathcal{M})$ be of the following form: $SS(\mathcal{M}) = \{t \cdot y \cdot (t-y) \cdot x = 0\}$. Denote the irreducible components $(S_1 : y = t)$ and $(\widetilde{S}_1 : x = 0)$.

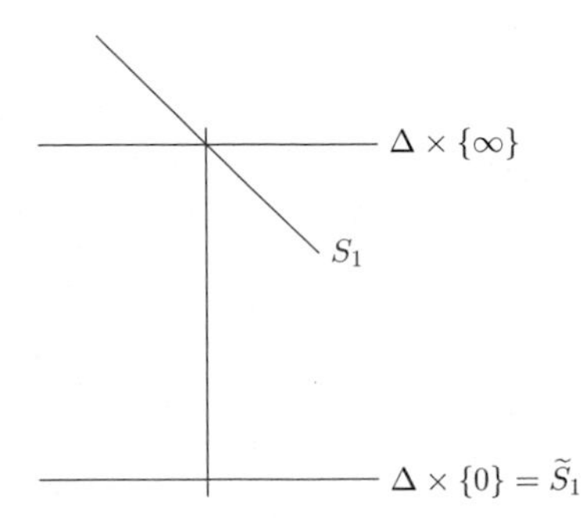

According to Theorem 3.1, $\hat{\mathcal{N}} = \mathcal{H}^0 p_+ (\mathcal{M} \otimes \mathcal{E}^q)_0^\wedge$ can be decomposed to:

$$\hat{\mathcal{N}} \cong R_0 \oplus \left(R_1 \otimes \mathcal{E}^{\frac{1}{t}}\right)$$

whereby

- $rk(R_0) = r$
- $rk(R_1) = r$

Moreover we have:

- $\dim\left(\mathcal{L}_{\leq 0}\right)_\vartheta = 2r$ for $\vartheta \in \left(\frac{\pi}{2}, \frac{3\pi}{2}\right)$ *mod* 2π
- $\dim\left(\mathcal{L}_{\leq 0}\right)_\vartheta = r$ for $\vartheta \in \left(-\frac{\pi}{2}, \frac{\pi}{2}\right)$ *mod* 2π
- $\dim\left(\mathcal{L}_{\leq \frac{1}{t}}\right)_\vartheta = 2r$ for $\vartheta \in \left(-\frac{\pi}{2}, \frac{\pi}{2}\right)$ *mod* 2π
- $\dim\left(\mathcal{L}_{\leq \frac{1}{t}}\right)_\vartheta = r$ for $\vartheta \in \left(\frac{\pi}{2}, \frac{3\pi}{2}\right)$ *mod* 2π

Remark 4.1: As we know from the Theorem of Levelt-Hukuhara-Turrittin the formal decomposition lifts to sectors of the real blow-up space. Let us take a closer look at the local system $\hat{\mathcal{L}}$ of $R_0 \oplus \left(R_1 \otimes \mathcal{E}^{\frac{1}{t}}\right)$. Obviously

$$\hat{\mathcal{L}} = \mathcal{R}_0 \oplus \mathcal{R}_1 e^{\frac{1}{t}}$$

(where $\mathcal{R}_i$ denote the local systems corresponding to the regular connections R_i). We can determine the stalks of $\hat{\mathcal{L}}_{\leq \psi}$ explicitly:

- $\vartheta \in \left(\frac{\pi}{2}, \frac{3\pi}{2}\right) : \left(\hat{\mathcal{L}}_{\leq 0}\right)_\vartheta = \left(\mathcal{R}_0 \oplus \mathcal{R}_1 e^{\frac{1}{t}}\right)_\vartheta = \hat{\mathcal{L}}_\vartheta,\ \left(\hat{\mathcal{L}}_{\leq \frac{1}{t}}\right)_\vartheta = \left(\mathcal{R}_1 e^{\frac{1}{t}}\right)_\vartheta$
- $\vartheta \in \left(-\frac{\pi}{2}, \frac{\pi}{2}\right) : \left(\hat{\mathcal{L}}_{\leq 0}\right)_\vartheta = \left(\mathcal{R}_0\right)_\vartheta,\ \left(\hat{\mathcal{L}}_{\leq \frac{1}{t}}\right)_\vartheta = \left(\mathcal{R}_0 \oplus \mathcal{R}_1 e^{\frac{1}{t}}\right)_\vartheta = \hat{\mathcal{L}}_\vartheta$

Thus we receive exhaustive filtrations of $\hat{\mathcal{L}}$ for every ϑ and therefore also for $\mathcal{L}$:

- $\vartheta \in \left(\frac{\pi}{2}, \frac{3\pi}{2}\right) : \left(\mathcal{L}_{\leq \frac{1}{t}}\right)_\vartheta \subset \left(\mathcal{L}_{\leq 0}\right)_\vartheta = \mathcal{L}_\vartheta$
- $\vartheta \in \left(-\frac{\pi}{2}, \frac{\pi}{2}\right) : \left(\mathcal{L}_{\leq 0}\right)_\vartheta \subset \left(\mathcal{L}_{\leq \frac{1}{t}}\right)_\vartheta = \mathcal{L}_\vartheta$

Now we will describe the Stokes structure via the isomorphism Ω (Theorem 3.8). Ω identifies

$$\left(\mathcal{L}_{\leq \psi}\right)_\vartheta \cong H^1\left(\left(\widetilde{p \circ e}\right)^{-1}(\vartheta), \beta^\vartheta_{\psi,!}\mathcal{F}_\psi\right).$$

Let us recapitulate the notations:

- $\mathcal{F}_\psi$ restriction of $\mathrm{DR}^{mod\, D_Z}\left(e^+\mathcal{M} \otimes \mathcal{E}^{g \circ e}\right)$ to its support $B^\vartheta_\psi \subset \left(\widetilde{p \circ e}\right)^{-1}$
- $\beta^\vartheta_\psi : B^\vartheta_\psi \hookrightarrow \left(\widetilde{p \circ e}\right)^{-1}(\vartheta)$ open inclusion
- $g(t,y) = \frac{1}{y} - \psi(t)$ for e a suitable sequence of point blow-ups

Lemma 4.2: *There exists a sequence of point blow-ups $e : Z \to X$, such that*

1. *the singular support of $\mathcal{M}$ becomes a normal crossing divisor*
2. *for both $\psi = 0$ and $\psi = \frac{1}{t}$ the exponent $g \circ e$ is holomorphic or good in every point.*

Proof: Consider the following blow-up of the divisor:

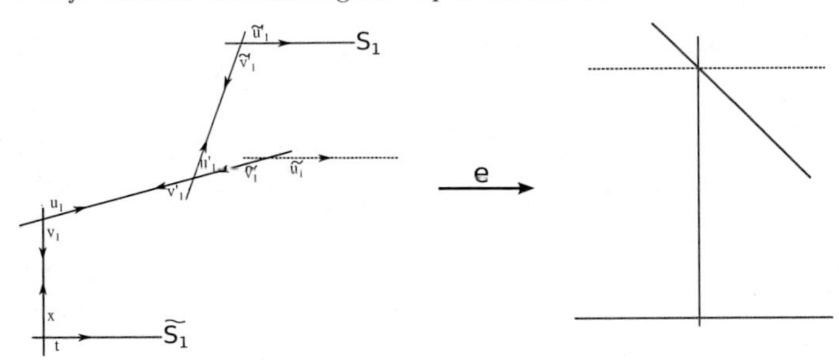

In local coordinates we get

1. for $\psi = 0$, i.e. $g(t, y) = \frac{1}{y}$:
   - $(g \circ e)(x, t) = x$ is holomorphic
   - $(g \circ e)(v_1, u_1) = \frac{1}{v_1}$ is good
   - $(g \circ e)(\tilde{u}_1, \tilde{v}_1) = \frac{1}{\tilde{u}_1 \tilde{v}_1}$ is good
   - $(g \circ e)(v'_1, u'_1) = \frac{1}{u'_1 v'_1 (v'_1 + 1)}$ is good
   - $(g \circ e)(\tilde{v}'_1, \tilde{u}'_1) = \frac{1}{\tilde{u}'_1 (\tilde{u}'_1 \tilde{v}'_1 + 1)}$ is good
2. for $\psi = \frac{1}{t}$, i.e. $g(t, y) = \frac{t-y}{ty}$:
   - $(g \circ e)(x, t) = \frac{tx-1}{t}$ is good
   - $(e \circ g)(u_1, v_1) = \frac{u_1 - 1}{u_1 v_1}$ is good for $u_1 \neq 1$
   - $(e \circ g)(\tilde{u}_1, \tilde{v}_1) = \frac{1-\tilde{v}_1}{\tilde{u}_1 \tilde{v}_1}$ is good for $\tilde{v}_1 \neq 1$
   - $(e \circ g)(u'_1, v'_1) = \frac{-1}{u'_1 (v'_1 + 1)}$ is good in $(0, 0)$ and holomorphic otherwise
   - $(e \circ g)(\tilde{u}'_1, \tilde{v}'_1) = \frac{-\tilde{v}'_1}{\tilde{u}'_1 \tilde{v}'_1 + 1}$ is holomorphic

□

We will denote the resulting normal crossing divisor by $\overline{D}_Z$. The fiber over ϑ with respect to the blow-up along $\overline{D}_Z$ is homeomorphic to a closed disc with two "holes". We will denote it by $\overline{A} \times \{\vartheta\}$.

Lemma 4.3: $\mathrm{DR}^{mod\,\overline{D}_Z}\left(e^+\mathcal{M} \otimes \mathcal{E}^{g\circ e}\right)$ *has cohomology in degree* 0 *at most.*

Proof: Since $\overline{D}_Z$ is normal crossing and $g \circ e$ is good or holomorphic (Lemma 4.2) the claim follows by [Sab13], Prop. 8.17 and Cor. 12.7.

□

Consider the map $\kappa : \widetilde{Z}\left(\overline{D}_Z\right) \to \widetilde{Z}\left(D_Z\right)$. Restricting it to a fiber $\kappa_\vartheta : \overline{A} \times \{\vartheta\} \to \left(\widetilde{p \circ e}\right)^{-1}(\vartheta)$, it is just the identity except at the points $\widetilde{P}_1$ and P_1, where it describes the "collapse" of the real blow-ups of $\widetilde{P}_1$ and P_1 back to these points. $\mathrm{DR}^{mod\,\overline{D}_Z}\left(e^+\mathcal{M} \otimes \mathcal{E}^{g\circ e}\right)$ has cohomology in degree 0 at most and therefore corresponds to a local system $\mathcal{G}_\psi$. Then $\mathrm{DR}^{mod\,D_Z}\left(e^+\mathcal{M} \otimes \mathcal{E}^{g\circ e}\right)$ corresponds to the perverse sheaf $\mathcal{F}_\psi$ given by $0 \to \kappa_*\mathcal{G}_\psi \to 0 \to 0$. Let $\mathcal{G}_\psi^\vartheta$ the restriction to the support $\kappa^{-1}\left(B_\psi^\vartheta\right)$ in the fiber $\overline{A} \times \{\vartheta\}$ and $\overline{\beta}_\psi^\vartheta : \kappa^{-1}\left(B_\psi^\vartheta\right) \hookrightarrow \overline{A} \times \{\vartheta\}$ the open inclusion. Because of Proposition 3.11 we have:

$$H^1\left(\overline{A} \times \{\vartheta\}, \overline{\beta}_{\psi,!}^\vartheta \mathcal{G}_\psi^\vartheta\right) \cong H^1\left(\left(\widetilde{p \circ e}\right)^{-1}(\vartheta), \beta_{\psi,!}^\vartheta \mathcal{F}_\psi^\vartheta\right)$$

Furthermore we denote by $\mathcal{K}$ the local system on $\widetilde{Z}\left(\overline{D}_Z\right)$ corresponding to the pullback connection $e^+\mathcal{M}$ of rank $rk(\mathcal{M})$ (see Cor. 8.3 in [Sab13]) and by $\mathcal{K}^\vartheta$ its restriction to $\overline{A} \times \{\vartheta\}$. Then obviously $\overline{\beta}_{\psi,!}^\vartheta \mathcal{G}_\psi^\vartheta$ equals $\overline{\beta}_{\psi,!}^\vartheta \overline{\beta}_\psi^{\vartheta\,-1} \mathcal{K}^\vartheta$ for all ψ. For brevity we will write $\overline{\beta}_{\psi,!}^\vartheta \mathcal{K}^\vartheta$ instead of $\overline{\beta}_{\psi,!}^\vartheta \overline{\beta}_\psi^{\vartheta\,-1} \mathcal{K}^\vartheta$ in the following.
Thus it is enough to examine:

$$\left(\mathcal{L}_{\leq\psi}\right)_\vartheta \cong H^1\left(\overline{A} \times \{\vartheta\}, \overline{\beta}_{\psi,!}^\vartheta \mathcal{K}^\vartheta\right)$$

With the blow-up e constructed in the proof of Lemma 4.2 we can determine the open subset $B_\vartheta^\psi \subset \left(\widetilde{p \circ e}\right)^{-1}(\vartheta)$:

1. $\psi = 0$:
   - $(0, \vartheta, |x|, \theta_x)$
   - $(0, \theta_{u_1}, |v_1|, \theta_{v_1}) \Leftrightarrow \theta_{v_1} \in \left(\frac{\pi}{2}, \frac{3\pi}{2}\right)$

- $\left(0, \vartheta, |v_{\bar{1}}|, \theta_{v_{\bar{1}}}\right) \Leftrightarrow \theta_{\tilde{v}_1} \in \left(\frac{\pi}{2} - \vartheta, \frac{3\pi}{2} - \vartheta\right)$
- $\left(|u'_1|, \theta_{u'_1}, 0, \theta_{v'_1}\right) \Leftrightarrow \vartheta \in \left(\frac{\pi}{2}, \frac{3\pi}{2}\right)$
- $\left(0, \vartheta, |v'_{\bar{1}}|, \theta_{v'_{\bar{1}}}\right) \Leftrightarrow \vartheta \in \left(\frac{\pi}{2}, \frac{3\pi}{2}\right)$

2. $\psi = \frac{1}{t}$
   - $(0, \vartheta, |x|, \theta_x) \Leftrightarrow \vartheta \in (-\frac{\pi}{2}, \frac{\pi}{2})$
   - $(0, \theta_{u_1}, |v_1|, \theta_{v_1}) \Leftrightarrow \vartheta \in (-\frac{\pi}{2}, \frac{\pi}{2})$
   - $\left(0, \vartheta, |v_{\bar{1}}|, \theta_{v_{\bar{1}}}\right) \Leftrightarrow arg\,(1 - v_{\bar{1}}) - \theta_{v_{\bar{1}}} \in \left(\frac{\pi}{2} + \vartheta, \frac{3\pi}{2} + \vartheta\right)$
   - $\left(|u'_1|, \theta_{u'_1}, 0, \theta_{v'_1}\right) \Leftrightarrow \theta_{u'_1} \in (-\frac{\pi}{2}, \frac{\pi}{2})$
   - $\left(0, \vartheta, |v'_{\bar{1}}|, \theta_{v'_{\bar{1}}}\right)$

We receive the following pictures, which show $\left(\widetilde{p \circ e}\right)^{-1}(\vartheta)$ (resp. $\overline{A} \times \{\vartheta\}$) and the subsets B_0^ϑ, $B_{\frac{1}{t}}^\vartheta$. One can see very clearly that by passing the Stokes directions $\pm\frac{\pi}{2}$, the relation of the subsets B_0^ϑ and $B_{\frac{1}{t}}^\vartheta$ changes from $B_0^\vartheta \subset B_{\frac{1}{t}}^\vartheta$ to $B_{\frac{1}{t}}^\vartheta \subset B_0^\vartheta$ and vice versa.

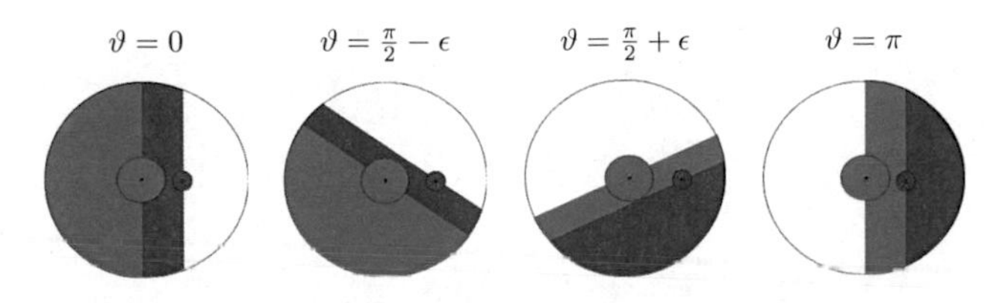

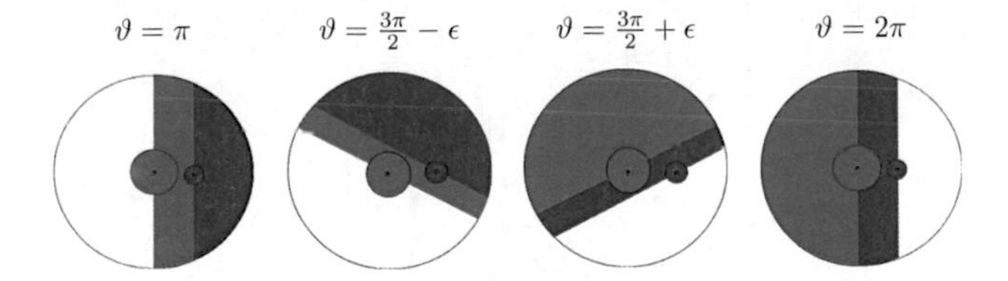

In this situation we can compute the first cohomology groups $H^1\left(\overline{A}\times\{\vartheta\},\overline{\beta}^{\vartheta}_{\psi,!}\mathcal{K}^{\vartheta}\right)$ in another way, namely by using Čech cohomology.

Lemma 4.4: *For every pair of angles $(\vartheta_0,\vartheta_1:=\vartheta_0+\pi)$ of $\mathbb{S}^1$ there exists a closed covering $\mathfrak{A}$ of $\overline{A}$, such that $\mathfrak{A}$ is a common Leray covering of $\beta^{\vartheta_i}_{\psi,!}\mathcal{K}^{\vartheta_i}$ ($i=1,2$ and $\psi=0,\frac{1}{t}$). Consequently we get an isomorphism*

$$H^1\left(\overline{A}\times\{\vartheta_i\},\overline{\beta}^{\vartheta_i}_{\psi,!}\mathcal{K}^{\vartheta_i}\right)\to \check{H}^1\left(\mathfrak{A},\overline{\beta}^{\vartheta_i}_{\psi,!}\mathcal{K}^{\vartheta_i}\right)$$

Proof: For $\vartheta_0\in[0,\frac{\pi}{2})$ consider the following curves $\alpha_1,\alpha_2,\alpha_3,\alpha_4$ in $\overline{A}$:

- α_1: line connecting the points $x_1:=\left(0,\frac{3\pi}{4}-2\vartheta_0\right)$ (in coordinates $(|x|,\theta_x)$) and $\left(0,-\frac{3\pi}{4}+\vartheta_0\right)$ (in coordinates $\left(|\widetilde{v}_1|,\theta_{\widetilde{v}_1}\right)$)
- α_2: line connecting the points $x_2:=(0,\frac{\pi}{4}-2\vartheta_0)$ (in coordinates $(|x|,\theta_x)$) and $(0,-\frac{\pi}{4}+\vartheta_0)$ (in coordinates $\left(|\widetilde{v}_1|,\theta_{\widetilde{v}_1}\right)$)
- α_3: line connecting the points $x_3:=\left(0,\frac{3\pi}{4}+\vartheta_0\right)$ (in coordinates $|\widetilde{v}'_1|,\theta_{\widetilde{v}'_1}$) and $\left(0,\frac{3\pi}{4}+\vartheta\right)$ (in coordinates $\left(|\widetilde{v}_1|,\theta_{\widetilde{v}_1}\right)$)
- α_4: line connecting the points $x_4:=(0,\frac{\pi}{4}+\vartheta_0)$ (in coordinates $|\widetilde{v}'_1|,\theta_{\widetilde{v}'_1}$) and $(0,\frac{\pi}{4}+\vartheta_0)$ (in coordinates $\left(|\widetilde{v}_1|,\theta_{\widetilde{v}_1}\right)$)

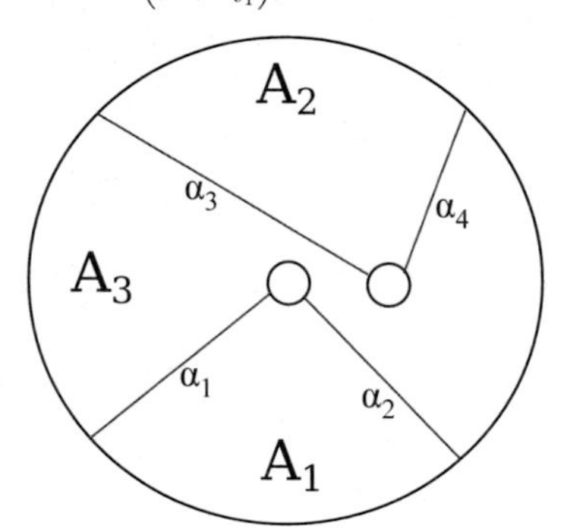

Then the closed covering $\mathfrak{A}=A_1\cup A_2\cup A_3$ of $\overline{A}$ defines a Leray covering of $\beta^{\vartheta_0}_{\psi,!}\mathcal{K}^{\vartheta_0}$ and $\beta^{\vartheta_1}_{\psi,!}\mathcal{K}^{\vartheta_1}$: This can be proved easily by using the exact sequence of Chapter 3.3.4 and doing the same calculations for the restriction to the intersections α_k.

For $\vartheta_0 \in [\frac{\pi}{2}, \pi)$ the following curves define a Leray covering for $\beta^{\vartheta_0}_{\psi,!}\mathcal{K}^{\vartheta_0}$ and $\beta^{\vartheta_1}_{\psi,!}\mathcal{K}^{\vartheta_1}$:

- $\alpha'_1 = \alpha_1$
- α'_2: line connecting the points $x'_2 := \left(0, -\frac{3\pi}{4} - 2\vartheta_0\right)$ (in coordinates $(|x|, \theta_x)$) and $\left(0, \frac{3\pi}{4} + \vartheta_0\right)$ (in coordinates $\left(|\widetilde{v}_1|, \theta_{\widetilde{v}_1}\right)$)
- α_3: line connecting the points $x'_3 := (0, -\frac{\pi}{4} + \vartheta_0)$ (in coordinates $|\widetilde{v}'_1|, \theta_{\widetilde{v}'_1}$) and $(0, -\frac{\pi}{4} + \vartheta)$ (in coordinates $\left(|\widetilde{v}_1|, \theta_{\widetilde{v}_1}\right)$)
- $\alpha'_4 = \alpha_4$

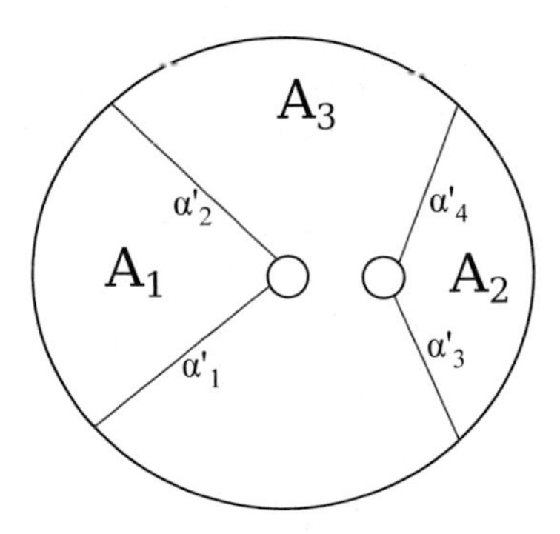

□

In the following lemma we will compute the cohomology groups concretely for $\vartheta = 0$ and $\vartheta = \pi$.

Lemma 4.5:

- $H^1\left(\overline{A} \times \{0\}, \overline{\beta}^0_{0,!}\mathcal{K}^0\right) \cong \mathcal{K}^0_{x_1}$
- $H^1\left(\overline{A} \times \{0\}, \overline{\beta}^{\frac{1}{t}}_{0,!}\mathcal{K}^0\right) = H^1\left(\overline{A} \times \{0\}, \overline{\beta}^0_!\mathcal{K}^0\right) \cong \mathcal{K}^0_{x_1} \oplus \mathcal{K}^0_{x_3}$
- $H^1\left(\overline{A} \times \{\pi\}, \overline{\beta}^{\frac{1}{t}}_{\pi,!}\mathcal{K}^\pi\right) \cong \mathcal{K}^\pi_{x_4}$
- $H^1\left(\overline{A} \times \{\pi\}, \overline{\beta}^0_{\pi,!}\mathcal{K}^\pi\right) = H^1\left(\overline{A} \times \{\pi\}, \overline{\beta}^\pi_!\mathcal{K}^\pi\right) \cong \mathcal{K}^\pi_{x_2} \oplus \mathcal{K}^\pi_{x_4}$

Proof:

1. $\vartheta = 0$:
   - $\psi = 0$:
     - $\check{C}^0 = H^0\left(A_1, \overline{\beta}^0_{0,!}\mathcal{K}^0\right) \oplus H^0\left(A_2, \overline{\beta}^0_{0,!}\mathcal{K}^0\right) \oplus H^0\left(A_3, \overline{\beta}^0_{0,!}\mathcal{K}^0\right) = 0$ since $A_i \not\subset I^0_0$.
     - $\check{C}^1 = H^0\left(\alpha_1, \overline{\beta}^0_{0,!}\mathcal{K}^0\right) \cong \mathcal{K}^0_{x_1}$ since $\alpha_1 \subset I^0_0$ and $\alpha_2, \alpha_3, \alpha_4 \not\subset I^0_0$

     $\Rightarrow H^1\left(\overline{A} \times \{0\}, \overline{\beta}^0_{0,!}\mathcal{K}^0\right) \cong \mathcal{K}^0_{x_1}$
   - $\psi = \frac{1}{t}$:
     - $\check{C}^0 = 0$
     - $\check{C}^1 = H^0\left(\alpha_1, \overline{\beta}^{\frac{1}{t}}_{0,!}\mathcal{K}^0\right) \oplus H^0\left(\alpha_3, \overline{\beta}^{\frac{1}{t}}_{0,!}\mathcal{K}^0\right) \cong \mathcal{K}^0_{x_1} \oplus \mathcal{K}^0_{x_3}$

     $\Rightarrow H^1\left(\overline{A} \times \{0\}, \overline{\beta}^{\frac{1}{t}}_{0,!}\mathcal{K}^0\right)\left(= H^1\left(\overline{A} \times \{0\}, \overline{\beta}^0_!\mathcal{K}^0\right)\right) \cong \mathcal{K}^0_{x_1} \oplus \mathcal{K}^0_{x_3}$
2. $\vartheta = \pi$:
   - $\psi = 0$:
     - $\check{C}^0 = 0$
     - $\check{C}^1 = H^0\left(\alpha_2, \overline{\beta}^0_{\pi,!}\mathcal{K}^\pi\right) \oplus H^0\left(\alpha_4, \overline{\beta}^0_{\pi,!}\mathcal{K}^\pi\right) \cong \mathcal{K}^\pi_{x_2} \oplus \mathcal{K}^\pi_{x_4}$

     $\Rightarrow H^1\left(\overline{A} \times \{\pi\}, \overline{\beta}^0_{\pi,!}\mathcal{K}^\pi\right)\left(= H^1\left(\overline{A} \times \{\pi\}, \overline{\beta}^\pi_!\mathcal{K}^\pi\right)\right) \cong \mathcal{K}^\pi_{x_2} \oplus \mathcal{K}^\pi_{x_4}$
   - $\psi = \frac{1}{t}$:
     - $\check{C}^0 = 0$
     - $\check{C}^1 = H^0\left(\alpha_4, \overline{\beta}^{\frac{1}{t}}_{\pi,!}\mathcal{K}^\pi\right) \cong \mathcal{K}^\pi_{x_4}$

     $\Rightarrow H^1\left(\overline{A} \times \{\pi\}, \overline{\beta}^{\frac{1}{t}}_{\pi,!}\mathcal{K}^\pi\right) \cong \mathcal{K}^\pi_{x_4}$

□

4.2 Stokes data associated to $\mathcal{L}$

In this section we develop a description of the Stokes-filtered local system $\mathcal{L}$ in terms of linear data, namely a set of linear Stokes data of exponential type.

Let $\Phi = \{\phi_i \mid i \in I\}$ denote a finite set of exponents ϕ_i of pole order ≤ 1 and let $\theta_0 \in \mathbb{S}^1$ be a generic angle, i. e. it is no Stokes direction with respect to the ϕ_is. We have already pointed out that we get a unique ordering of the exponents $\phi_0 <_{\theta_0} \phi_1 <_{\theta_0} \cdots <_{\theta_0} \phi_n$ and the reversed ordering for $\theta_1 := \theta_0 + \pi$.

Definition 4.6: The *category of Stokes data of exponential type* (for a set of exponents Φ ordered by θ_0) has objects consisting of two families of $\mathbb{C}$-vector spaces

$$(G_{\phi_i}, H_{\phi_i})$$

and two morphisms

$$\bigoplus_{i=0}^{n} G_{\phi_i} \xrightarrow{S} \bigoplus_{i=0}^{n} H_{\phi_i} \qquad \bigoplus_{i=0}^{n} H_{\phi_{n-i}} \xrightarrow{S'} \bigoplus_{i=0}^{n} G_{\phi_{n-i}}$$

such that

1. S is a block upper triangular matrix, i. e. $S_{ij} : G_{\phi_i} \to H_{\phi_j}$ is zero for $i > j$ and S_{ii} is invertible (thus S is invertible and $\dim G_{\phi_i} = \dim H_{\phi_i}$)
2. S' is a block lower triangular matrix, i. e. $S'_{ij} : H_{\phi_{n-i}} \to G_{\phi_{n-j}}$ is zero for $i < j$ and S'_{ii} is invertible (thus S' is invertible)

A morphism consists of morphisms of $\mathbb{C}$-vector spaces $\lambda_i^G : G_{\phi_i} \to G'_{\phi_i}$ and $\lambda_i^H : H_{\phi_i} \to H'_{\phi_i}$, which are compatible with the corresponding diagrams.

The correspondence between Stokes-filtered local systems of exponential type and linear Stokes data is stated in the following

Theorem 4.7: *There is an equivalence of categories between the Stokes-filtered local systems of exponential type of Definition 3.3 and Stokes data of exponential type of Definition 4.6.*

Proof: We refer to [HS11], p. 12/13. □

Using the functor constructed in [HS11] we can associate a set of Stokes data to our Stokes-filtered local system $\mathcal{L}$ described above (Remark 4.1).

Construction 4.8: Fix two intervals

$$I_0 = (0 - \epsilon, \pi + \epsilon), \quad I_1 = (-\pi - \epsilon, 0 + \epsilon)$$

of length $\pi + 2\epsilon$ on $\mathbb{S}^1$, such that the intersection $I_0 \cap I_1$ consists of $(0 - \epsilon, 0 + \epsilon)$ and $(\pi - \epsilon, \pi + \epsilon)$. Observe that the intersections do not contain the Stokes directions $-\frac{\pi}{2}, \frac{\pi}{2}$. To our given local system $\mathcal{L}$ we associate:

- a vector space associated to the angle 0, i. e. the stalk $\mathcal{L}_0$. It comes equipped with the Stokes filtration.
- a vector space associated to the angle π, i. e. the stalk $\mathcal{L}_\pi$, coming equipped with the Stokes filtration.
- vector spaces associated to the intervals I_0 and I_1, i. e. the global sections $\Gamma(I_0, \mathcal{L})$, $\Gamma(I_1, \mathcal{L})$
- a diagram of isomorphisms (given by the natural restriction to the stalks):

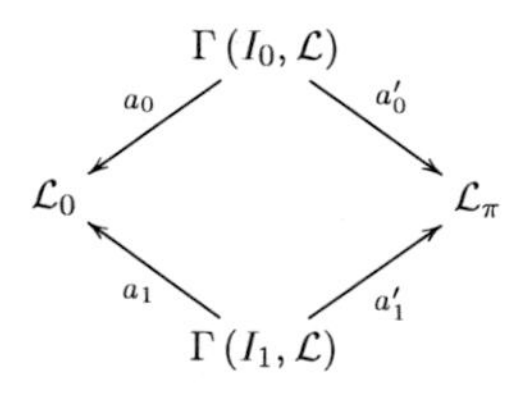

The filtrations on the stalks are opposite with respect to the maps $a_0' a_0^{-1}$ and $a_1 a'^{-1}_1$, i. e.

$$\mathcal{L}_0 = \bigoplus_{\phi \in \{0, \frac{1}{t}\}} L_{\leq \phi, 0} \cap a_0' a_0^{-1} (L_{\leq \phi, \pi}), \qquad \mathcal{L}_\pi = \bigoplus_{\phi \in \{0, \frac{1}{t}\}} L_{\leq \phi, \pi} \cap a_1 a'^{-1}_1 (L_{\leq \phi, 0})$$

Furthermore we know, by using the isomorphisms Ω and Γ, that $\mathcal{L}_0 \cong \mathcal{K}_{x_1} \oplus \mathcal{K}_{x_3}$ whereby $\mathcal{L}_{\leq 0,0} \cong \mathcal{K}_{x_1}$. Thus we have a splitting $\mathcal{L}_0 = G_0 \oplus G_{\frac{1}{t}}$ with $G_0 = \mathcal{L}_{\leq 0,0}$. The same holds for $\mathcal{L}_\pi$: $\mathcal{L}_\pi = H_0 \oplus H_{\frac{1}{t}}$ with $H_{\frac{1}{t}} = \mathcal{L}_{\leq \frac{1}{t}, \pi}$.

Thus the set of data $\left(G_0, G_{\frac{1}{t}}, H_0, H_{\frac{1}{t}}, S_0^\pi, S_\pi^0\right)$ with:

- $\mathcal{L}_0 = G_0 \oplus G_{\frac{1}{t}}$ and $\mathcal{L}_\pi = H_{\frac{1}{t}} \oplus H_0$
- $S_0^\pi : G_0 \oplus G_{\frac{1}{t}} \xrightarrow{a_0^{-1}} \Gamma(I_0, \mathcal{L}) \xrightarrow{a_0'} H_0 \oplus H_{\frac{1}{t}}$
- $S_\pi^0 : H_{\frac{1}{t}} \oplus H_0 \xrightarrow{a_1^{-1}} \Gamma(I_1, \mathcal{L}) \xrightarrow{a_1'} G_{\frac{1}{t}} \oplus G_0$

describes a set of Stokes data associated to the local system $\mathcal{L}$.

By exhaustivity of the filtrations on $\mathcal{L}_\vartheta$ ($\vartheta = 0, \pi$) the isomorphism Ω induces

$$\mathcal{L}_\vartheta \cong H^1\left(\overline{A} \times \vartheta, \overline{\beta}_!^\vartheta \mathcal{K}^\vartheta\right)$$

where $\overline{\beta}^\vartheta : B^\vartheta \hookrightarrow \overline{A} \times \vartheta$ corresponds to the $\overline{\beta}_\psi^\vartheta$ with $\psi \geq_\vartheta \phi$ for $\psi, \phi \in \Phi = \{0, \frac{1}{t}\}$ and $B^\vartheta := B_\psi^\vartheta \supset B_\phi^\vartheta$.
In the same way, for an open interval $I \subset \mathbb{S}^1$ let $\mathcal{K}^I$ be the restriction of $\mathcal{K}$ to $\overline{A} \times I$ and define $\overline{\beta}^I : B^I \hookrightarrow \overline{A} \times I$ the inclusion of the subspace B^I, which is the support of $\mathrm{DR}^{mod\ D}(\mathcal{M} \otimes \mathcal{E}^{\frac{1}{y}})$ according to Remark 3.28. Notice that $\overline{\beta}^I_{|\overline{A} \times \vartheta} = \overline{\beta}^\vartheta$. Then, by the isomorphism Ω, we identify

$$\Gamma(I, \mathcal{L}) \cong H^1\left(\overline{A} \times I, \overline{\beta}_!^I \mathcal{K}^I\right).$$

With these isomorphisms we have restriction morphisms (according to the restrictions to the stalks)

$$\rho_\vartheta : H^1\left(\overline{A} \times I_0, \overline{\beta}_!^{I_0} \mathcal{K}^{I_0}\right) \xrightarrow{\cong} H^1\left(\overline{A} \times \vartheta, \overline{\beta}_!^\vartheta \mathcal{K}^\vartheta\right) \text{ for } \vartheta \in I_0$$

$$\rho_\vartheta' : H^1\left(\overline{A} \times I_1, \overline{\beta}_!^{I_1} \mathcal{K}^{I_1}\right) \xrightarrow{\cong} H^1\left(\overline{A} \times \vartheta, \overline{\beta}_!^\vartheta \mathcal{K}^\vartheta\right) \text{ for } \vartheta \in I_1$$

This yields to a new way of describing Stokes data associated to the local system $\mathcal{L}$:

Theorem 4.9: *Set the following data:*

- *vector spaces* $L_0 := H^1\left(\overline{A} \times \{0\}, \overline{\beta}_!^0 \mathcal{K}^0\right)$ *and* $L_1 := H^1\left(\overline{A} \times \{\pi\}, \overline{\beta}_!^\pi \mathcal{K}^\pi\right)$
- *morphisms* $\sigma_0^\pi := \rho_\pi \circ \rho_0^{-1}$ *and* $\sigma_\pi^0 := \rho_0' \circ {\rho'}_\pi^{-1}$ *, with* ρ_ϑ *and* ρ_ϑ' *defined as above.*

Then

$$\left(L_0, L_1, \sigma_0^\pi, \sigma_\pi^0\right)$$

defines a set of Stokes data for $\mathcal{H}^0 p_+ \left(\mathcal{M} \otimes \mathcal{E}^{\frac{1}{y}}\right)$.

Proof: The isomorphism $\Omega : \mathcal{L}_\vartheta \to H^1\left(\overline{A} \times \{\vartheta\}, \overline{\beta}_!^\vartheta \mathcal{K}^\vartheta\right)$ passes the filtrations on $\mathcal{L}_0$ and $\mathcal{L}_\pi$ to L_0 and L_1 and therefore yields to a suitable graduation of $L_0 = \widetilde{G}_0 \oplus \widetilde{G}_{\frac{1}{t}} \cong \mathcal{K}_{x_1}^0 \oplus \mathcal{K}_{x_3}^0$ and $L_1 = \widetilde{H}_{\frac{1}{t}} \oplus \widetilde{H}_0 \cong \mathcal{K}_{x_4}^\pi \oplus \mathcal{K}_{x_2}^\pi$.
On the level of graduated spaces $\Omega : G_0 \oplus G_{\frac{1}{t}} \to \widetilde{G}_0 \oplus \widetilde{G}_{\frac{1}{t}}$ (resp. $H_{\frac{1}{t}} \oplus H_0 \to \widetilde{H}_{\frac{1}{t}} \oplus \widetilde{H}_0$) is obviously described by a block diagonal matrix. Furthermore, by definition of ρ and ρ', the maps σ_0^π and σ_π^0 can be read as $\sigma_0^\pi = \Omega \circ S_0^\pi \circ \Omega^{-1}$ and $\sigma_\pi^0 = \Omega \circ S_\pi^0 \circ \Omega^{-1}$. Since, according to the construction in 4.8, S_0^π is upper block triangular (resp. S_π^0 is lower block triangular) the same holds for σ_0^π (resp. σ_π^0).

□

4.3 Explicit computation of the Stokes matrices

The determination of the Stokes data thus corresponds to the following picture:

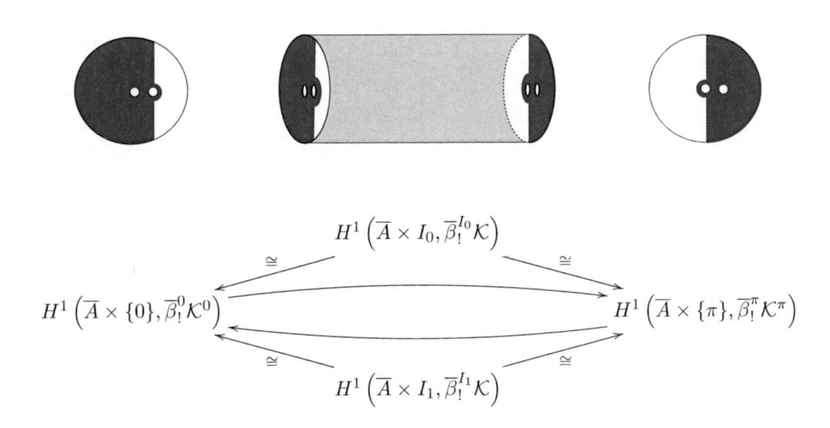

In Lemma 4.5 we have already computed the cohomology groups for $\vartheta = 0$ and $\vartheta = \pi$ using the Leray covering $\mathfrak{A}$. Now for $l = 0, 1$ we fix a diffeomorphism $\overline{A}\times I_l \xrightarrow{\sim} \overline{A}\times I_l$ by lifting the vector field ∂_ϑ to $\overline{A}\times\mathbb{S}^1$ such that the lift is equal to ∂_ϑ away from a small neighborhood of $\partial\overline{A}$ and such that the diffeomorphism induces $B^{I_l} \xrightarrow{\cong} B^{\vartheta_{l+1}}\times I_l$ where $\vartheta_1 = \pi$ and $\vartheta_2 = 0$. It induces a diffeomorphism $\overline{A}\times\{\vartheta_l\} \xrightarrow{\cong} \overline{A}\times\{\vartheta_{l+1}\}$ and an isomorphism between the push forward of $\mathcal{K}^{\vartheta_l}$ and $\mathcal{K}^{\vartheta_{l+1}}$. Moreover it sends the boundary ∂B^{ϑ_l} to $\partial B^{\vartheta_{l+1}}$ (i. e. the boundaries are rotated in counter clockwise direction by the angle π). Via this diffeomorphism the curves $\alpha_i \subset \overline{A}\times\vartheta_l$ are sent to curves $\widetilde{\alpha}_i \subset \overline{A}\times\vartheta_{l+1}$ and therefore induce another Leray covering $\widetilde{\mathfrak{A}}$ of $H^1\left(\overline{A}\times\{\vartheta_{l+1}\}, \overline{\beta}_!^{\vartheta_{l+1}}\mathcal{K}^{\vartheta_{l+1}}\right)$.

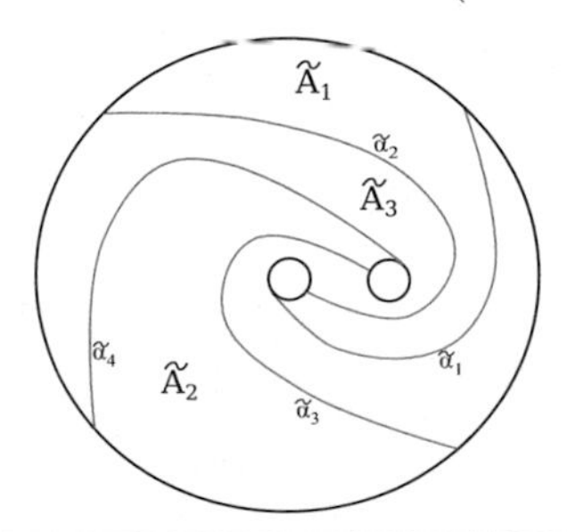

Explicitely, for $l = 0$ we get a Leray covering $\widetilde{\mathfrak{A}}$ of $H^1\left(\overline{A}\times\{\pi\},\overline{\beta}_!^\pi\mathcal{K}^\pi\right)$ and, as in the previous chapter, we get an isomorphism $\widetilde{\Gamma}_\pi : H^1\left(\overline{A}\times\{\pi\},\overline{\beta}_!^\pi\mathcal{K}^\pi\right)\to \check{H}^1\left(\widetilde{\mathfrak{A}},\overline{\beta}_!^\pi\mathcal{K}^\pi\right)$, which gives us:

$$H^1\left(\overline{A}\times\{\pi\},\overline{\beta}_!^\pi\mathcal{K}^\pi\right)\cong\mathcal{K}^\pi\left(\widetilde{\alpha}_1\right)\oplus\mathcal{K}^\pi\left(\widetilde{\alpha}_3\right)\cong\mathcal{K}^\pi_{x_1}\oplus\mathcal{K}^\pi_{x_3}$$

Thus the above diffeomorphism leads to an isomorphism

$$\mu_0^\pi : \mathcal{K}^0_{x_1}\oplus\mathcal{K}^0_{x_3}\xrightarrow{\cong}\mathcal{K}^\pi_{x_1}\oplus\mathcal{K}^\pi_{x_3}$$

Furthermore let us fix the following vector space $\mathbb{V} := \mathcal{K}^\pi_c$ to be the stalk of the local system $\mathcal{K}$ at the point $c := \left(0,\pi,\frac{1}{2},0\right)$ (which is obviously a point in the fiber $\overline{A}\times\{\pi\}$). By analytic continuation we can identify every non-zero stalk $\mathcal{K}^\vartheta_x$ with $\mathbb{V}$ for all ϑ.
Now consider the following diagram of isomorphisms:

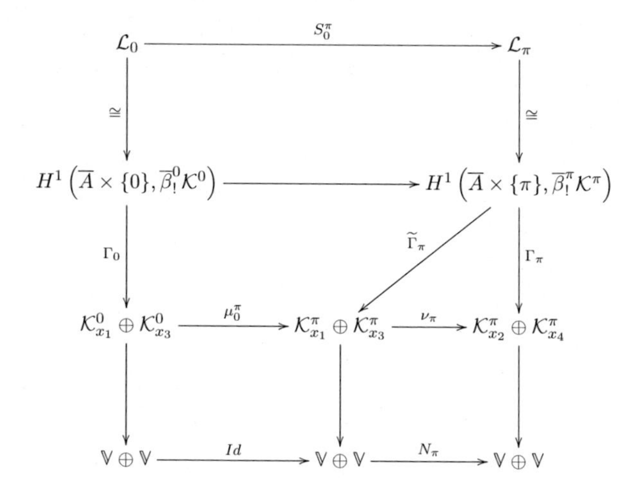

It remains to determine the map ν_π (respectively N_π). Therefore we will combine the coverings $\mathfrak{A}$ and $\widetilde{\mathfrak{A}}$ of $\overline{A}\times\{\pi\}$ to a refined covering $\mathfrak{B}$. We get refinement maps $ref_{\mathfrak{A}\to\mathfrak{B}}$ and $ref_{\widetilde{\mathfrak{A}}\to\mathfrak{B}}$ and receive the following picture:

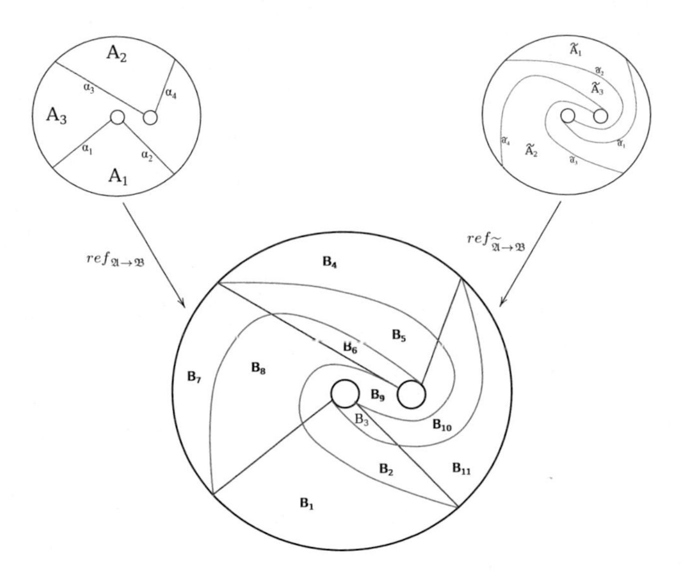

Since $\mathfrak{B}$ is again a Leray covering the refinement maps induce isomorphisms on the cohomology groups. Thus we will 'extend' the above diagram of isomorphisms in the following way:

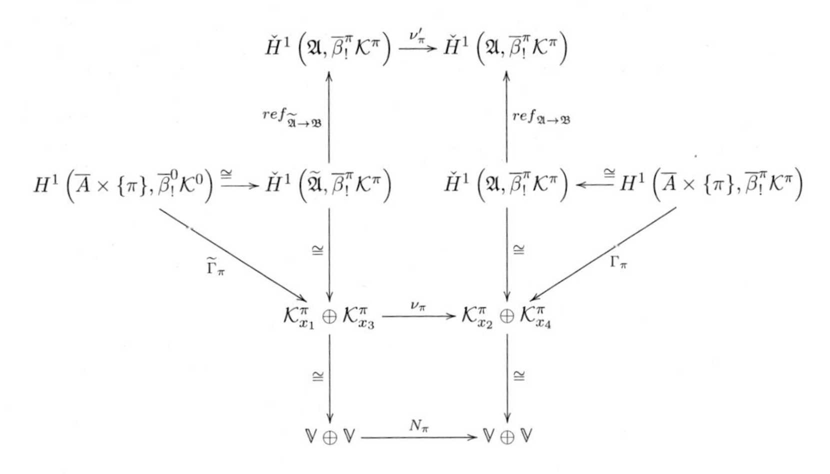

To determine the refinement maps on the first cohomology groups we have to consider the Čech complex for $\mathfrak{B}$.

We set the following index sets

- $I := \{1, 2, 3\}$ (indices corresponding to the covering $\mathfrak{A} = \bigcup_{i \in I} A_i$)
- $\widetilde{I} := \{1, 2, 3\}$ (corresponding to $\widetilde{\mathfrak{A}} = \bigcup_{i \in \widetilde{I}} \widetilde{A}_i$)
- $J := \{1, 2, 3, \ldots, 11\}$ (corresponding to $\mathfrak{B} = \bigcup_{j \in J} B_j$)
- $K := I \times J$
- $J' := \{2, 3, 9, 10, 11\} \subset J$
- $K' = \{(1,2), (1,9), (1,11), (2,3), (2,8), (2,9), (2,10), (2,11), (3,9), (3,10), (3,11), (4,9), (4,10), (4,11), (5,9), (5,10), (6,9), (8,9), (9,10), (10,11)\} \subset K$

The Čech complex for $\mathfrak{B}$ is given by:

$$\begin{array}{ccccc} \check{C}^0 & \xrightarrow{d_0} & \check{C}^1 & \xrightarrow{d_1} & \check{C}^2 \xrightarrow{d_2} \cdots \\ \| & & \| & & \\ \bigoplus_{j \in J'} \check{H}^0\left(B_j, \overline{\beta}_!^{\pi} \mathcal{K}^{\pi}\right) & & \bigoplus_{(i,j) \in K'} \check{H}^0\left(B_i \cup B_j, \overline{\beta}_!^{\pi} \mathcal{K}^{\pi}\right) & & \end{array}$$

To determine the map d_0 we will use the following identification of $\check{C}^1$:
As above, by analytic continuation we can identify each component $\check{H}^0\left(B_i \cup B_j, \overline{\beta}_!^{\pi} \mathcal{K}^{\pi}\right)$ with $\mathbb{V}^{k(i,j)} := \bigoplus_{l=1}^{k(i,j)} \mathbb{V}$ where $k(i,j)$ denotes the number of connected components of $(B_i \cap B_j)$. $k(i,j) = 1$ for all $(i,j) \in J$ except for $(3,9)$ and $(6,9)$ where it is equal to 2. So we have an isomorphism

$$\check{C}^1 \cong \bigoplus_{(i,j) \in K'} \mathbb{V}^{k(i,j)}$$

Now if we take a section b_9 of B_9, restrict it to a boundary component of B_9 (which is the intersection with one of the bordering B_is) and identify it with $\mathbb{V}$, we have to take care of the monodromies S, T around the two leaks in $\overline{A} \times \{\pi\}$. The following picture shows, by restriction to which boundary component we receive monodromy.

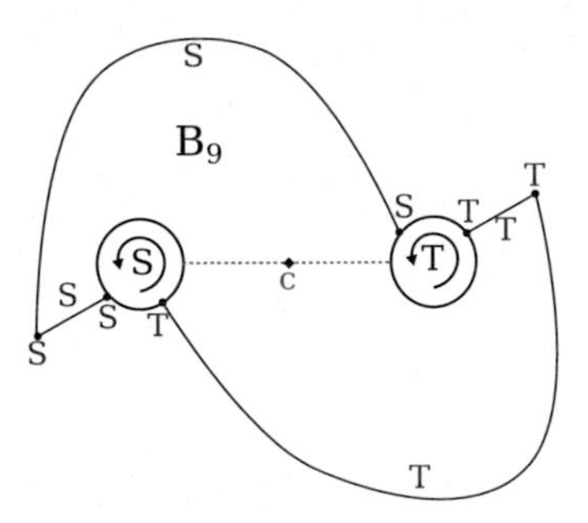

Remark 4.10: If we follow S and T in the coordinates of $\overline{A} \times \{\vartheta\}$, we can also describe them in terms of monodromy around the divisor components: S can be described by a path $\gamma_S : [0,1] \to \overline{A} \times \{\vartheta\}, \tau \mapsto (0, \vartheta, 0, \tau \cdot 2\pi)$ in the coordinates $(|t|, \vartheta, |x|, \theta_x)$. If we look at this path in the (complex) blow-up of the singular locus of $\mathcal{M}$, γ_S corresponds to the monodromy around $\widetilde{S}_1$. In the same way T is given by $\gamma_T : [0,1] \to \overline{A} \times \{\vartheta\}, \tau \mapsto (0, \vartheta, 0, \tau \cdot 2\pi)$ in the coordinates $\left(|\tilde{u}_1'|, \vartheta, |\tilde{v}_1'|, \theta_{\tilde{v}_1'}\right)$ and it corresponds to the monodromy around the strict transform of S_1. Note that S and T do not depend on ϑ.

The map d_0 is given by:

$$\begin{array}{ccc} \check{C}^0 & \xrightarrow{d_0} & \check{C}^1 \cong \bigoplus_{(i,j)\in K'} \mathbb{V}^{k(i,j)} \\ (b_2, b_3, b_9, b_{10}, b_{11}) & \longmapsto & (b_2, Sb_9, b_{11}, b_3 - b_2, -b_2, Sb_9 - b_2, b_{10} - b_2, b_{11} - b_2, \\ & & Sb_9 - b_3, Tb_9 - b_3, b_{10} - b_3, b_{11} - b_3, Tb_9, b_{10}, b_{11}, \\ & & Tb_9, b_{10}, Sb_9, Tb_9, Sb_9, b_{10} - Tb_9, b_{11} - b_{10}) \end{array}$$

The refinement map $ref_{\mathfrak{A}\to\mathfrak{B}} : \check{H}^1\left(\mathfrak{A}, \overline{\beta}_!^\pi \mathcal{K}^\pi\right) \to \check{H}^1\left(\mathfrak{B}, \overline{\beta}_!^\pi \mathcal{K}^\pi\right)$ is determined by the index map

$$\begin{array}{ccc} J & \longrightarrow & I \\ 1,2,3 & \longmapsto & 1 \\ 4,5,6 & \longmapsto & 2 \\ 7,8,9,10,11 & \longmapsto & 3 \end{array}$$

and we obtain:

$$\begin{array}{cccc} \check{H}^1\left(\mathfrak{A},\overline{\beta_!^\pi}\mathcal{K}^\pi\right)\cong\mathcal{K}^\pi_{x_2}\oplus\mathcal{K}^\pi_{x_4} & \overset{ref_{\mathfrak{A}\to\mathfrak{B}}}{\longrightarrow} & \check{H}^1\left(\mathfrak{B},\overline{\beta_!^\pi}\mathcal{K}^\pi\right) & \\ (a_2,a_4) & \mapsto & (0,0,a_2,0,0,0,a_2,a_2, & \\ & & a_2,a_2,a_2,a_4,a_4,a_4, & \mod im(d_0) \\ & & a_4,a_4,a_4,0,0,0) & \end{array}$$

In the same way the refinement map $ref_{\widetilde{\mathfrak{A}}\to\mathfrak{B}} : \check{H}^1\left(\widetilde{\mathfrak{A}},\overline{\beta_!^\pi}\mathcal{K}^\pi\right)\to\check{H}^1\left(\mathfrak{B},\overline{\beta_!^\pi}\mathcal{K}^\pi\right)$ is determined by the index map

$$\begin{array}{ccc} J & \longrightarrow & \widetilde{I} \\ 3,4,10 & \longmapsto & 1 \\ 1,6,8 & \longmapsto & 2 \\ 2,5,7,9,11 & \longmapsto & 3 \end{array}$$

and we obtain:

$$\begin{array}{cccc} \check{H}^1\left(\widetilde{\mathfrak{A}},\overline{\beta_!^\pi}\mathcal{K}^\pi\right)\cong\mathcal{K}^\pi_{x_1}\oplus\mathcal{K}^\pi_{x_3} & \overset{ref_{\widetilde{\mathfrak{A}}\to\mathfrak{B}}}{\longrightarrow} & \check{H}^1\left(\mathfrak{B},\overline{\beta_!^\pi}\mathcal{K}^\pi\right) & \\ (\tilde{a}_1,\tilde{a}_3) & \mapsto & (\tilde{a}_3,\tilde{a}_3,\tilde{a}_3,-\tilde{a}_1,-\tilde{a}_3,0, & \\ & & -\tilde{a}_1,0,\tilde{a}_1,0,\tilde{a}_1,0,0, & \mod im\,(d_0) \\ & & \tilde{a}_1,0,0,\tilde{a}_3,\tilde{a}_3,0,\tilde{a}_1) & \end{array}$$

Now let (a_2,a_4) be a base of $\check{H}^1\left(\mathfrak{A},\overline{\beta_!^\pi}\mathcal{K}^\pi\right)\cong\mathcal{K}^\pi_{x_2}\oplus\mathcal{K}^\pi_{x_4}\cong\mathbb{V}\oplus\mathbb{V}$ and $(\tilde{a}_1,\tilde{a}_3)$ a base of $\check{H}^1\left(\widetilde{\mathfrak{A}},\overline{\beta_!^\pi}\mathcal{K}^\pi\right)\cong\mathcal{K}^\pi_{x_1}\oplus\mathcal{K}^\pi_{x_3}\cong\mathbb{V}\oplus\mathbb{V}$. We receive two bases of $\check{H}^1\left(\mathfrak{B},\overline{\beta_!^\pi}\mathcal{K}^\pi\right)$, namely $ref_{\widetilde{\mathfrak{A}}\to\mathfrak{B}}\,(\tilde{a}_1,\tilde{a}_3)$ and $ref_{\mathfrak{A}\to\mathfrak{B}}\,(a_2,a_4)$ and thus ν' is determined by representing the base $ref_{\widetilde{\mathfrak{A}}\to\mathfrak{B}}\,(\tilde{a}_1,\tilde{a}_3)$ in terms of $ref_{\mathfrak{A}\to\mathfrak{B}}\,(a_2,a_4)$.
As before we identify $\check{C}^1\left(\mathfrak{B},\overline{\beta_!^\pi}\mathcal{K}^\pi\right)\cong\bigoplus\mathbb{V}^{k(i,j)}$ and we end up in solving the following equation for each $(i,j)\in K'$:

$$\left(ref_{\widetilde{\mathfrak{A}}\to\mathfrak{B}}\,(\tilde{a}_1,\tilde{a}_3)\right)_{(i,j)}=\left(ref_{\mathfrak{A}\to\mathfrak{B}}\,(a_2,a_4)\right)_{(i,j)}\mod im\,(d_0)$$

Explicitely:

$$
\begin{array}{ll}
(1,2): & \tilde{a}_3 = 0 + b_2 \\
(1,9): & \tilde{a}_3 = 0 + Sb_9 \\
(1,11): & \tilde{a}_3 = a_2 + b_{11} \\
(2,3): & -\tilde{a}_1 = 0 + b_3 - b_2 \\
(2,8): & -\tilde{a}_3 = -b_2 \\
(2,9): & 0 = 0 + Sb_9 - b_2 \\
(2,10): & -\tilde{a}_1 = a_2 + b_{10} - b_2 \\
(2,11): & 0 = a_2 + b_{11} - b_2 \\
(3,9): & \tilde{a}_1 = 0 + Sb_9 - b_3,\ 0 = a_2 + Tb_9 - b_3 \\
(3,10): & 0 = a_2 + b_{10} - b_3 \\
(3,11): & \tilde{a}_1 = a_2 + b_{11} - b_3 \\
(4,9): & 0 = a_4 + Tb_9 \\
(4,10): & 0 = a_4 + b_{10} \\
(4,11): & \tilde{a}_1 = a_4 + b_{11} \\
(5,9): & 0 = a_4 + Tb_9 \\
(5,10): & 0 = a_4 + b_{10} \\
(6,9): & \tilde{a}_3 = 0 + Sb_9,\ 0 = a_4 + Tb_9 \\
(8,9): & \tilde{a}_3 = 0 + Sb_9 \\
(9,10): & 0 = 0 + b_{10} - Tb_9 \\
(10,11): & \tilde{a}_1 = 0 + b_{11} - b_{10}
\end{array}
$$

We get the following result:

$$\tilde{a}_1 = -a_2 + \left(1 - ST^{-1}\right) a_4, \ \ \tilde{a}_3 = -ST^{-1} a_4$$

(whereby $b_2 = -ST^{-1}a_4, b_3 = a_2 - a_4, b_9 = -T^{-1}a_4, b_{10} = -a_4, b_{11} = -ST^{-1}a_4 - a_2$) and consequently the map N_π is given by the matrix

$$\begin{pmatrix} -1 & 1 - ST^{-1} \\ 0 & -ST^{-1} \end{pmatrix}$$

For calculating $S_\pi^0 : \mathcal{L}_\pi \to \mathcal{L}_0$, we will use exactly the same procedure, except that we have to take care about the continuation to the vector space $\mathbb{V}$.
First we fix another vector space $\mathbb{W} := \mathcal{K}_c^0$ where $c = \left(0, 0, \frac{1}{2}, 0\right) \in \overline{A} \times \{0\}$ and consider the following diagram:

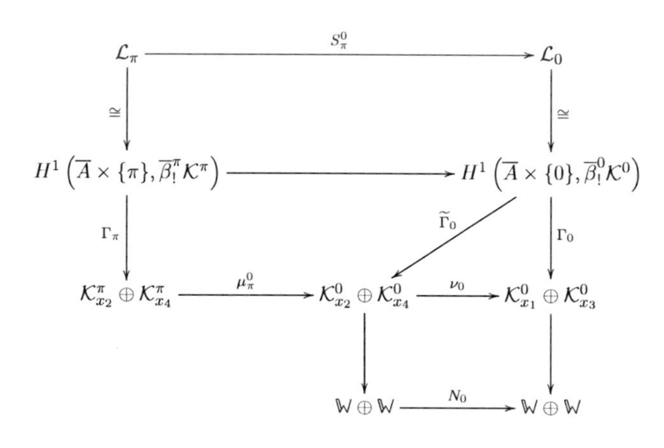

As before, for the determination of the map ν_0 (respectively N_0) we combine the coverings $\mathfrak{A}$ and $\widetilde{\mathfrak{A}}$ of $\overline{A} \times \{0\}$ to the refined covering $\mathfrak{B}$ and get the refinement maps $ref_{\mathfrak{A}\to\mathfrak{B}}$ and $ref_{\widetilde{\mathfrak{A}}\to\mathfrak{B}}$ which induce isomorphisms on the cohomology groups.

We set the following index sets

- $I := \{1, 2, 3\}$ (indices corresponding to the covering $\mathfrak{A} = \bigcup_{i\in I} A_i$)
- $\widetilde{I} := \{1, 2, 3\}$ (corresponding to $\widetilde{\mathfrak{A}} = \bigcup_{i\in \widetilde{I}} \widetilde{A}_i$)
- $J := \{1, 2, 3, \ldots, 11\}$ (corresponding to $\mathfrak{B} = \bigcup_{j\in J} B_j$)
- $K := I \times J$
- $J' := \{5, 6, 7, 8, 9\} \subset J$
- $K' = \{(1,7), (1,8), (1,9), (2,8), (2,9), (3,9), (4,5), (4,7), (5,6), (5,7), (5,8), (5,9), (5,10), (6,7), (6,8), (6,9), (7,8), (8,9), (9,10)\} \subset K$

The map d_0 is given by:

$$\begin{array}{ccc} \check{C}^0 & \xrightarrow{d_0} & \check{C}^1 \cong \bigoplus_{(i,j)\in K'} \mathbb{W}^{k(i,j)} \\ (b_5,b_6,b_7,b_8,b_9) & \longmapsto & (b_7,b_8,Sb_9,b_8,Sb_9,Sb_9-b_3,Tb_9-b_3,b_5,b_7, \\ & & b_6-b_5,b_7-b_5,b_8-b_5,Tb_9-b_5,-b_5,b_7-b_6, \\ & & b_8-b_6,Sb_9-b_6,Tb_9-b_6,b_8-b_7,Sb_9-b_8,-Tb_9) \end{array}$$

The refinement maps are given by:

$$\begin{array}{cccc} \check{H}^1\left(\mathfrak{A},\overline{\beta}_!^0\mathcal{K}^0\right) \cong \mathcal{K}^0_{x_1}\oplus\mathcal{K}^0_{x_3} & \xrightarrow{ref_{\mathfrak{A}\to\mathfrak{B}}} & \check{H}^1\left(\mathfrak{B},\overline{\beta}_!^0\mathcal{K}^0\right) & \\ (a_1,a_3) & \mapsto & (a_1,a_1,a_1,a_1,a_1,a_1, & \\ & & 0,0,a_3,0,a_3,a_3,0,0, & \bmod im(d_0) \\ & & a_3,a_3,a_3,0,0,0,0) & \end{array}$$

and

$$\begin{array}{cccc} \check{H}^1\left(\widetilde{\mathfrak{A}},\overline{\beta}_!^0\mathcal{K}^0\right) \cong \mathcal{K}^0_{x_2}\oplus\mathcal{K}^0_{x_4} & \xrightarrow{ref_{\widetilde{\mathfrak{A}}\to\mathfrak{B}}} & \check{H}^1\left(\mathfrak{B},\overline{\beta}_!^0\mathcal{K}^0\right) & \\ (\tilde{a}_2,\tilde{a}_4) & \mapsto & (\tilde{a}_4,0,0,0,0,0,\tilde{a}_2,\tilde{a}_2, & \\ & & \tilde{a}_2,-\tilde{a}_4,0,-\tilde{a}_4,0,-\tilde{a}_2, & \bmod im(d_0) \\ & & \tilde{a}_4,0,0,\tilde{a}_4,-\tilde{a}_4,0,-\tilde{a}_2) & \end{array}$$

The system of equations

$$\begin{array}{ll} (1,7) & \tilde{a}_4 = a_1 + b_7 \\ (1,8) & 0 = a_1 + b_8 \\ (1,9) & 0 = a_1 + Sb_9 \\ (2,8) & 0 = a_1 + b_8 \\ (2,9) & 0 = a_1 + Sb_9 \\ (3,9) & 0 = a_1 + Sb_9 - b_3,\ \tilde{a}_2 = 0 + Tb_9 - b_3 \\ (4,5) & \tilde{a}_2 = 0 + b_5 \\ (4,7) & \tilde{a}_2 = a_3 + b_7 \\ (5,6) & -\tilde{a}_4 = 0 + b_6 - b_5 \end{array}$$

$$\begin{array}{ll}
(5,7) & 0 = a_3 + b_7 - b_5 \\
(5,8) & -\tilde{a}_4 = a_3 + b_8 - b_5 \\
(5,9) & 0 = 0 + Tb_9 - b_5 \\
(5,10) & -\tilde{a}_2 = 0 - b_5 \\
(6,7) & \tilde{a}_4 = a_3 + b_7 - b_6 \\
(6,8) & 0 = a_3 + b_8 - b_6 \\
(6,9) & 0 = a_3 + Sb_9 - b_6,\ \tilde{a}_4 = 0 + Tb_9 - b_6 \\
(7,8) & -\tilde{a}_4 = 0 + b_8 - b_7 \\
(8,9) & 0 = 0 + Sb_9 - b_8 \\
(9,10) & -\tilde{a}_2 = 0 - Tb_9
\end{array}$$

is solved by

$$\tilde{a}_2 = -TS^{-1}a_1,\ \ \tilde{a}_4 = \left(1 - TS^{-1}\right) a_1 - a_3$$

(whereby $b_5 = -TS^{-1}a_1, b_6 = a_3 - a_1, b_7 = -TS^{-1}a_1 - a_3, b_8 = -a_1, b_9 = -S^{-1}a_1$) and thus N_0 is given by the matrix

$$\begin{pmatrix} -TS^{-1} & 0 \\ 1 - TS^{-1} & -1 \end{pmatrix}$$

We extend ν_0 respectively N_0 to the vector space $\mathbb{V} \oplus \mathbb{V}$ by μ_π^0 (which does not affect N_0):

$$\begin{array}{ccccccc}
\mathcal{K}_{x_2}^{\pi} \oplus \mathcal{K}_{x_4}^{\pi} & \xrightarrow{\mu_\pi^0} & \mathcal{K}_{x_2}^{0} \oplus \mathcal{K}_{x_4}^{0} & \xrightarrow{\nu_0} & \mathcal{K}_{x_1}^{0} \oplus \mathcal{K}_{x_3}^{0} & \xleftarrow{\mu_\pi^0} & \mathcal{K}_{x_1}^{\pi} \oplus \mathcal{K}_{x_3}^{\pi} \\
\downarrow & & \downarrow & & \downarrow & & \downarrow \\
\mathbb{V} \oplus \mathbb{V} & \xrightarrow{Id} & \mathbb{W} \oplus \mathbb{W} & \xrightarrow{N_0} & \mathbb{W} \oplus \mathbb{W} & \xleftarrow{Id} & \mathbb{V} \oplus \mathbb{V}
\end{array}$$

Summarizing all the calculated isomorphisms we receive the following diagram:

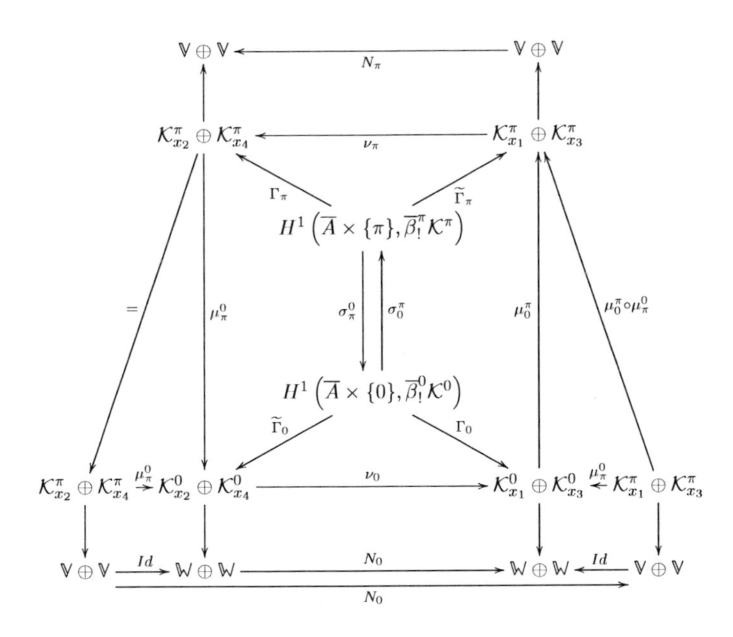

Let us fix two isomorphisms Σ_0 and Σ_π, which we will call the standard identification of $H^1\left(\overline{A} \times \{0\}, \overline{\beta}^0_! \mathcal{K}^0\right)$, respectively $H^1\left(\overline{A} \times \{\pi\}, \overline{\beta}^\pi_! \mathcal{K}^\pi\right)$ with the vector space $\mathbb{V} \oplus \mathbb{V}$.

$$\Sigma_0 : H^1\left(\overline{A} \times \{0\}, \overline{\beta}^0_! \mathcal{K}^0\right) \xrightarrow{\Gamma_0} \mathcal{K}^0_{x_1} \oplus \mathcal{K}^0_{x_3} \xrightarrow{\mu^\pi_0} \mathcal{K}^\pi_{x_1} \oplus \mathcal{K}^\pi_{x_3} \to \mathbb{V} \oplus \mathbb{V}$$

$$\Sigma_\pi : H^1\left(\overline{A} \times \{\pi\}, \overline{\beta}^\pi_! \mathcal{K}^\pi\right) \xrightarrow{\Gamma_\pi} \mathcal{K}^\pi_{x_2} \oplus \mathcal{K}^\pi_{x_4} \to \mathbb{V} \oplus \mathbb{V}$$

Summarizing the previous calculations, we can state the following theorem. It provides an explicit way of computing Stokes matrices for the direct image $\mathcal{H}^0 p_+ \left(\mathcal{M} \otimes \mathcal{E}^{\frac{1}{y}}\right)$.

Theorem 4.11: *Fix the following data:*

- $L_0 := \mathbb{V} \oplus \mathbb{V},\ L_1 := \mathbb{V} \oplus \mathbb{V}$
- $S_0^1 = N_\pi = \begin{pmatrix} -1 & 1 - ST^{-1} \\ 0 & -ST^{-1} \end{pmatrix},\ S_1^0 = (\mu_0^\pi \circ \mu_\pi^0) \cdot N_0 = \begin{pmatrix} U & 0 \\ 0 & U \end{pmatrix} \cdot \begin{pmatrix} -TS^{-1} & 0 \\ 1 - TS^{-1} & -1 \end{pmatrix}$

where S and T denote the monodromies around the strict transforms of the irreducible components $\widetilde{S}_1$ and S_1 in the singular locus of $\mathcal{M}$ and U denotes the mondromy around the component $\{0\} \times \mathbb{P}^1$. Then

$$\left(L_0, L_1, S_0^1, S_1^0\right)$$

defines a set of Stokes data for $\mathcal{H}^0 p_+ \left(\mathcal{M} \otimes \mathcal{E}^{\frac{1}{y}}\right)$.

Proof: At first remark that

$$\mu_0^\pi \circ \mu_\pi^0 : \mathcal{K}_x^\pi \oplus \mathcal{K}_y^\pi \to \mathcal{K}_x^0 \oplus \mathcal{K}_y^0 \to \mathcal{K}_x^\pi \oplus \mathcal{K}_y^\pi$$

is the isomorphism arising from varying the angel ϑ via the path

$$\gamma_U : [0,1] \to \mathbb{S}^1, \tau \mapsto \pi + \tau \cdot 2\pi.$$

This corresponds to the monodromy U around the divisor component $\{0\} \times \mathbb{P}^1$. Furthermore from Theorem 4.9 we know that

$$\left(H^1\left(\overline{A} \times \{0\}, \beta_!^0 \mathcal{K}^0\right), H^1\left(\overline{A} \times \{\pi\}, \beta_!^\pi \mathcal{K}^\pi\right), \sigma_0^\pi, \sigma_\pi^0\right)$$

defines a set of Stokes data. With the standard identifications of our vector spaces we can rewrite the above diagram to:

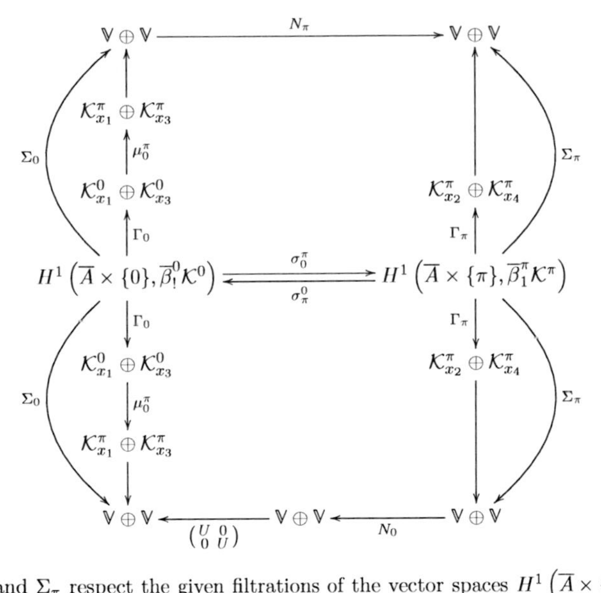

Since Σ_0 and Σ_π respect the given filtrations of the vector spaces $H^1\left(\overline{A}\times\{0\},\overline{\beta}^0_!\mathcal{K}^0\right)$ and $H^1\left(\overline{A}\times\{\pi\},\overline{\beta}^\pi_1\mathcal{K}^\pi\right)$, it follows that the induced filtrations on $\mathbb{V}\oplus\mathbb{V}$ are mutually opposite with respect to $S^1_0=\Sigma_\pi\circ\sigma^\pi_0\circ{\Sigma_0}^{-1}$ and $S^0_1=\Sigma_0\circ\sigma^0_\pi\circ{\Sigma_\pi}^{-1}$.

□

Bibliography

[Dim04] Alexandru Dimca. *Sheaves in topology*. Springer-Verlag, Berlin, 2004.

[DK] Andrea D'Agnolo and Masaki Kashiwara. Riemann-hilbert correspondence for holonomic $\mathcal{D}$-modules. http://arxiv.org/pdf/1311.2374v1.pdf.

[HS] Marco Hien and Claude Sabbah. The local laplace transform of an elementary irregular meromorphic connection. http://arxiv.org/pdf/1405.5310v1.pdf.

[HS11] Claus Hertling and Claude Sabbah. Examples of non-commutative hodge structures. *J. Inst. Math. Jussieu*, 10(3):635–674, 2011.

[Ive86] Birger Iversen. *Cohomology of sheaves*. Springer-Verlag, Berlin, 1986.

[Ked10] Kiran S. Kedlaya. Good formal structures for flat meromorphic connections, I: surfaces. *Duke Math. J.*, 154(2):343–418, 2010.

[Mal91] Bernard Malgrange. *Equations Différentielles à Coefficients Polynomiaux*, volume 96 of *Progress in Mathematics*. Birkhäuser, 1991.

[Meb89] Zoghman Mebkhout. *Le formalisme des six opérations de Grothendieck pour les $\mathcal{D}_X$-modules cohérents*, volume 35 of *Travaux en Cours [Works in Progress]*. Hermann, Paris, 1989.

[Moc] Takuro Mochizuki. Holonomic $\mathcal{D}$-modules with betti structure. http://arxiv.org/pdf/1001.2336v5.pdf.

[Moc09] Takuro Mochizuki. Good formal structure for meromorphic flat connections on smooth projective surfaces. In *Algebraic analysis and around*, volume 54 of *Adv. Stud. Pure Math.*, pages 223–253. Math. Soc. Japan, Tokyo, 2009.

[Rou07] Céline Roucairol. Formal structure of direct image of holonomic $\mathcal{D}$-modules of exponential typed-modules of exponential type. *Manuscripta Math.*, 124(3):299–318, 2007.

[Sab07] Claude Sabbah. *Isomonodromic deformations and Frobenius manifolds.* Springer, London, 2007.

[Sab08] Claude Sabbah. An explicit stationary phase formula for the local formal Fourier-Laplace transform. In *Singularities I*, volume 474 of *Contemp. Math.*, pages 309–330. Amer. Math. Soc., Providence, RI, 2008.

[Sab13] Claude Sabbah. *Introduction to Stokes structures*, volume 2060 of *Lecture Notes in Mathematics.* Springer, Heidelberg, 2013.

Lebenslauf

Name	Hedwig Heizinger
Geburtsdaten	7. September 1984 in Landshut
Schulausbildung	Gymnasium Seligenthal, Landshut (1994-2003)
Hochschulausbildung	Studium der Musikwissenschaft, Soziologie, Psychologie (Magister Artium) und der Mathematik (Diplom) an der Universität Regensburg (2003-2011)
Promotion	Universität Augsburg (2011-2015)
Kontakt	Universität Augsburg Lehrstuhl für Algebra und Zahlentheorie D-86153 Augsburg